Model-based Design and Optimization of
Vanadium Redox Flow Batteries

„Model-based Design and Optimization of Vanadium Redox Flow Batteries"

Zur Erlangung des akademischen Grades eines

DOKTOR-INGENIEURS

von der Fakultät für
Elektrotechnik und Informationstechnik
des Karlsruher Instituts für Technologie
genehmigte

DISSERTATION

von

Dipl.-Ing. *Sebastian König*
geb. in: Saarlouis, Deutschland

Tag der mündlichen Prüfung:	19. Juli 2017
Hauptreferent:	Prof. Dr.-Ing. T. Leibfried
Korreferent:	Prof. Dr. rer. nat. J. Tübke

Bibliografische Information der Deutschen Nationalbibliothek

Die Deutsche Nationalbibliothek verzeichnet diese Publikation in der Deutschen Nationalbibliografie; detaillierte bibliografische Daten sind im Internet über http://dnb.d-nb.de abrufbar.

1. Aufl. - Göttingen: Cuvillier, 2017
  Zugl.: Karlsruhe (KIT), Univ., Diss., 2017

© CUVILLIER VERLAG, Göttingen 2017
  Nonnenstieg 8, 37075 Göttingen
  Telefon: 0551-54724-0
  Telefax: 0551-54724-21
  www.cuvillier.de

  ISBN 978-3-7369-9615-1
  eISBN 978-3-7369-8615-2

"It thus can be appreciated that designing a flow battery to maximize system energy efficiency is a somewhat complex process requiring analytical models, experimental data and iterative solutions"

Final Report of the NASA Redox Storage System Development Project in 1984 [1].

Danksagung (Acknowledgements)

Die vorliegende Arbeit entstand während meiner Tätigkeit als wissenschaftlicher Mitarbeiter am Institut für Elektroenergiesysteme und Hochspannungstechnik (IEH) des Karlsruher Instituts für Technologie (KIT). Für die vielfältige Unterstützung in dieser Zeit möchte ich mich bei vielen Menschen bedanken.

Herrn Prof. Thomas Leibfried danke ich vor allem für sein kontinuierliches Interesse an der Thematik und die gewährte Freiheit zur selbständigen Ausrichtung des Forschungsthemas. Auch für die Übernahme des Hauptreferats und das gute Feedback bezüglich der schriftlichen Ausarbeitung gebührt im Dank.
Herrn Prof. Jens Tübke danke ich für die Übernahme des Korreferats, sein andauerndes Interesse an meiner Arbeit und die vielen guten Diskussionen.

Vielen Dank an Christine Eisinger und Isabell Riedmüller für die vielfältig und immer bereitwillig gewährte Unterstützung in allen verwaltungstechnischen Belangen.

Bei der RWE Power AG bedanke ich mich für die Aufnahme in das Stipendienprogramm „Power Engineers" und die daraus resultierende Förderung und Unterstützung während meines Studiums. Hier möchte ich insbesondere meinen Mentor Herrn Dr. Jörg Walter dankend hervorheben, der für mich auch über die Zeit des Stipendienprogramms hinaus ein wertvoller Ansprechpartner geblieben ist.

Den Firmen SCHMID Energy Systems, J. Schmalz und Gildemeister Energy Storage danke ich für die Bereitstellung der experimentellen Daten. In diesem Kontext bedanke ich mich vor allem bei Ruben Wößner für die unkomplizierte Kooperationsbereitschaft. Großer Dank gebührt auch Peter Fischer vom Fraunhofer ICT für viele Stunden fruchtbarer Diskussion und die gewährte Nachhilfe in Elektrochemie.

Für die geleistete Arbeit bedanke ich mich weiterhin bei allen von mir betreuten Studierenden und Hiwis, allen voran Ann-Kathrin Brenner, Arne Wöber und Martin Zimmerlin.

Für die gewährte gute Freundschaft, die stetige Motivation zu sportlicher Aktivität und die unzähligen wertvollen fachlichen und nicht-fachlichen Diskussionen gilt Martin Uhrig mein besonderer Dank.
Meinen früheren Kommilitonen, heutigen Arbeitskollegen und guten Freunden Simon Wenig, Nico Meyer-Hübner und Yannick Rink danke ich für die vielen nicht immer ganz ernsten Diskussionen und Fachsimpeleien, die die Zeit am IEH so besonders gemacht haben.

Großer Dank gebührt auch Thomas Lüth, der in seiner damaligen Rolle als Referent der Geschäftsführung von TRUMPF Hüttinger den Kontakt zur Firma Gebr. SCHMID herstellte und mir so die sinnvolle Fortsetzung meiner Arbeit ermöglichte.

Mein tief empfundener Dank gilt meinen Eltern, Maria und Wolfgang König, die mein technisches Interesse stets gefördert und mir das Studium ermöglicht haben. Vielen Dank für eure Unterstützung, auf die ich immer zählen kann.
Meinen Schwiegereltern Doris Schorr-Eisenbeis und Gisbert Eisenbeis danke ich herzlich für ihre großartige Unterstützung besonders während des letzten Jahres meiner Promotion.

Der allergrößte Dank aber gebührt meiner geliebten Ehefrau Nina-Kathrin. Ohne ihren Rückhalt und ihre unermüdliche Unterstützung wäre die vorliegende Arbeit niemals entstanden.

Karlsruhe, im August 2017

Sebastian König

Contents

Zusammenfassung (German abstract)

Die vorliegende Arbeit besteht aus drei Teilen. Unter zu Hilfenahme zahlreicher Publikationen wird zuerst ein umfangreiches multiphysikalisches Modell einer Vanadium Redox Flow Batterie (VRFB) aufgebaut. Das Model nutzt den Ansatz der konzentrierten Parameter um die Batterie auf der Systemebene zu beschreiben. Die unterschiedlichen Teilmodelle werden durch numerische Beispiele illustriert.

Dank der Unterstützung durch drei verschiedene Hersteller kann das erstellte Modell einer umfangreichen Validierung unterzogen werden. Die weithin verwendeten Literaturmodelle zeigen sich nicht in der Lage, die Realität mit akzeptabler Genauigkeit abzubilden. Daher wird mit Hilfe der Messdaten der Hersteller ein Versuch unternommen, die elektrochemisch aktive Oberfläche des Graphitfilzes, welcher als Elektrode dient, zu bestimmen. Diese Oberfläche ist ein kritischer Eingabeparameter für das Modell der Konzentrationspolarisation. Nach der Validierung und unter Einsatz der neu bestimmten Fläche ist das Modell in der Lage, die Systeme zweier Hersteller mit hoher Genauigkeit zu beschreiben. Auch das System des dritten Herstellers wird wesentlich genauer beschrieben als vom Literaturmodell, jedoch verbleiben hier noch signifikante Abweichungen.

Im zweiten Teil der Arbeit wird das validierte Modell genutzt um eine umfangreiche Designstudie durchzuführen. Mit Hilfe der Finite-Element-Analyse werden insgesamt 24 verschiedene Zellentwürfe erstellt, die sich in der verwendeten Elektrodenfläche und dem Aufbau der internen Elektrolytzu- und -ableitung unterscheiden. Die entworfenen Zellen werden anschließend in Stacks auf ihre Eignung in einer VRFB mit nur einem Stack und einer VRFB mit einem Strang aus drei seriell verschalteten Stacks hin untersucht. Ein Alleinstellungsmerkmal der durchgeführten Studie ist die gleichzeitig erfolgende Optimierung des Volumenstroms, welche für eine bestmögliche Vergleichbarkeit der verschiedenen Entwürfe unerlässlich ist.

Anhand der Studie können vier wichtige Zusammenhänge identifiziert werden:

- Für die Reduktion der Streuströme ist eine Vergrößerung der Elektrodenfläche und ein längerer und engerer Zuleitungskanal gleich effektiv. Beide Maßnahmen erfahren jedoch einen starken Sättigungseffekt.

- Für einen gegebenen Flussfaktor und ein konstantes Seitenverhältnis erzielt eine Elektrode mit einer größeren Fläche eine höhere Strömungsgeschwindigkeit des Elektrolyten und daher auch einen höheren Massentransferkoeffizienten. Umgekehrt benötigt eine größere Elektrode für einen optimalen Betrieb einen kleineren Flussfaktor.

- Bei einer großen Elektrode verursacht ein Zuleitungskanal mit einem vernünftigen Geometriefaktor einen überproportional hohen Druckabfall verglichen mit kleineren Elektroden, bedingt durch den großen Elektrolytbedarf.

- Die Serienschaltung mehrerer Stacks erhöht zwar die Batteriespannung, führt jedoch zu einem niedrigeren Systemwirkungsgrad. Abhängig von der verwendeten Stromdichte liegt dieser Verlust zwischen 1,6 und 3,3 Prozentpunkten. Für höhere Stromdichten scheint es jedoch möglich, den Verlust an Wirkungsgrad auszugleichen durch einen Batteriewechselrichter der dank der höheren Eingangsspannung selbst einen höheren Wirkungsgrad hat.

Für die Batterie mit nur einem Stack, welches folgerichtig alle Lastfälle bedienen muss, wurde eine Zelle mit einer 2000 cm^2 großen Elektrode und einem langen aber nicht zu engen Zuleitungskanal als effizientester Entwurf identifiziert. Mit kleinen Abstrichen beim Wirkungsgrad ist es jedoch möglich, die Elektrodenfläche zwischen 1000 cm^2 und 4000 cm^2 weitgehend frei zu wählen.

In der Serienschaltung der Stacks treten deutlich erhöhte Streuströme auf. Der Vorteil der größeren Elektrode bezüglich der Streuströme wird teilweise kompensiert durch die Notwendigkeit, die großen Zellen mit großen Rohr- und Schlauchdurchmessern zu versorgen. Trotzdem erzielt die größte untersuchte Zelle mit dem längsten und engsten Zuleitungskanal den höchsten Wirkungsgrad in einem Strang aus drei Stacks. Für einen solchen Strang muss von der Verwendung einer kleinen Elektrodenfläche in Kombination mit einem kurzen und breiten Zuleitungskanal abgeraten werden.

Insgesamt gewinnt das Thema Zelldesign an Bedeutung, wenn mehrere Stacks elektrisch in Serie betrieben werden sollen.

Im dritten und letzten Teil der Arbeit wird die bereits vorab veröffentlichte innovative Volumenstromregelung präsentiert und erweitert. Ein Spannungsregler für die Stackspannung wird vorgestellt, welcher über die Regelung des Volumenstroms die Verletzung von Zellspannungsgrenzen so lange wie möglich verzögert. Dieser Regler ist in der Lage die Effizienz der innovativen Volumenstromregelung mit einer hohen Entladekapazität zu verbinden.

Betrachtet man ausschließlich den Wirkungsgrad, erscheint die Verwendung eines konstanten Volumenstroms als Option. Jedoch muss mit dieser ein deutlicher Verlust an Entladekapazität akzeptiert werden. Im Vergleich mit der konventionellen variablen Volumenstromregelung nach Faradays erstem Gesetz der Elektrolyse erreicht der innovative Ansatz eine Steigerung des Wirkungsgrads um einen Prozentpunkt.

Für die nominale Stromdichte erscheint die Verwendung eines konstanten Volumenstroms auch unter Berücksichtigung einer maximalen Entladekapazität möglich. Für kleinere Stromdichten muss dann jedoch ein Wirkungsgradverlust von bis zu 21 Prozentpunkten in Kauf genommen werden. Die innovative variable Volumenstromregelung erzielt leicht höhere Kapazitäten als die konventionelle. Gleichzeitig erreicht sie zudem eine Steigerung des Wirkungsgrads um bis zu 0,9 Prozentpunkte.

In Kombination mit dem vorgestellten Spannungsregler ist die innovative variable Volumenstromregelung daher allen anderen Konzepten überlegen.

Chapter 1

Introduction and motivation

1.1 The need for electric energy storage

In 2015, every German grid-connected consumer had access to electricity with an average interruption (SAIDI – system average interruption index) of 12.7 min [2]. This corresponds to an outstanding supply security of 99.9998 %. The requirement of a secure electricity supply seems to contradict utilizing renewable energy sources (RES), such as photovoltaics (PV) and wind power plants because of their volatile nature. However, RES apparently had a positive effect on the supply security, as shown in Figure 1-1. Between 2006 and 2015, the share of RES in the German gross electricity consumption nearly tripled, while the SAIDI was cut in half.

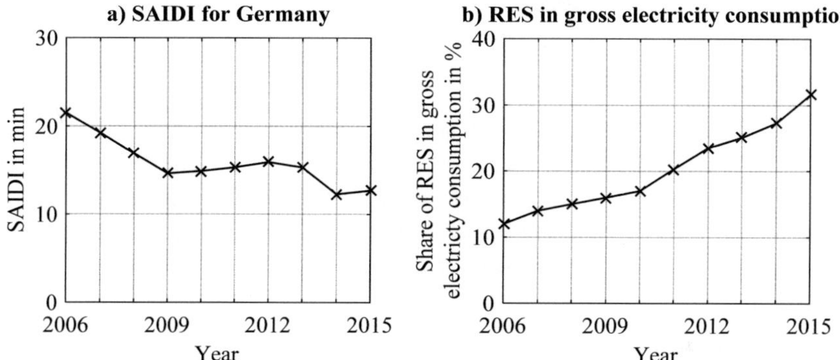

Figure 1-1: SAIDI and share of RES in the gross electricity consumption in Germany

It is beyond dispute that the de-carbonization of an energy system requires additional flexibility options, to balance generation and demand. Besides energy storage systems (EES), wide-area grid expansion and demand side management (DSM) will also contribute to these flexibilities.

Among all energy storage technologies, batteries are the most flexible but also the most expensive one. In March 2017, the global energy storage database of the U.S. Department of Energy listed 323 grid-scale battery storage projects with an individual nominal power of 1 MW or more. This number includes all contracted projects, projects under construction and projects in operation. In total, the database lists battery energy storage systems (BESS) with a total power of 2,804 MW and a total capacity of 4,176 MWh. For comparison, pumped-storage power plants with a total power of 6,850 MW are installed in Germany only [3]. Hence today, batteries play a minor role in grid-scale energy storage.

However, in 2016, several major BESS projects were published. Three of them, with a total power of 341 MW, are given here as a sample. National Grid, the transport system operator (TSO) of England and Wales, procured 201 MW of so-called enhanced frequency response (EFR) contracts after a tendering exercise [4]. All accepted bids went to providers that plan to install BESS.

In Southern California, we can find another recent example for the large-scale penetration of the electric power grid by batteries. In 2015, a massive leak in the Aliso Canyon storage facility, a natural gas storage plant, caused an ongoing natural gas shortage [5]. As this fuel is in particular used by the so-called 'peaker plants' – gas-driven power plants that deliver energy in times of peaking demand, the region currently faces a significant lack of fast responding generation units [6]. Therefore, the utilities Southern California Edison and San Diego Gas & Electric procured more than 50 MW of lithium-ion batteries to manage this shortage. The largest project within this tender is a 20 MW/80 MWh battery system.

Also, between mid-2016 and the beginning of 2017, the German energy company steag GmbH installed six 15 MW/22 MWh batteries in their power plants in Germany [7]. The company markets these systems to deliver primary frequency reserve power, but they add flexibility to steag's power plant portfolio also.

Because of versatile marketing options, short installation times and falling prices, the market for grid-scale BESS is growing dynamically. While today, batteries first and foremost represent high-power ESS, there will also be a need of high-energy ESS in the future, e.g., to replace fossil-fueled peaker plants and to store PV energy for the nighttime.

1.2 Fundamentals

1.2.1 Setup and characteristics of a flow battery

A flow battery comprises the same elements as a conventional battery. It consists of two electrodes, in conventional batteries often denoted as anode (negative electrode) and cathode (positive electrode), a separator to isolate both electrodes, plus an electrolyte, which enables an ionic charge transfer between the electrodes. However, in flow batteries, the tasks of these elements are different. First and foremost, the electrodes in a flow battery do not partake in the redox reactions. They just offer the surface area for the electrochemical reaction to take place, and the electric conductivity to distribute and to collect the electrons. Thus, the electrodes of a flow battery do not store any energy. Instead, the energy storage is an additional task of the electrolyte. Therefore, we have to separate the electrolyte into the negative electrolyte (anolyte) and the positive electrolyte (catholyte). An ion-exchange membrane prevents the two electrolytes in the flow cell from mixing.

According to the technical definition, the denotation anode and cathode, as well as anolyte and catholyte, refer to the discharging process. During the discharging process, the cathode is the positive pole, and thus absorbs electrons; the anode is the negative pole and thus releases electrons. During the charging process, the polarity of the battery cell does not change, but the roles of the electrodes do. The negative electrode now has to absorb electrons and thus it becomes the cathode; the positive electrode becomes the anode. Hence, in a battery, the denotations anode and cathode are conflicting. Consequentially, the terms positive and negative electrode, as well as positive and negative electrolyte are exclusively used in this work.

The deployment of liquid electrolytes for storing the energy, offers a key advantage. By replacing the electrolyte via a pair of inlets and outlets, the redox flow cell can be used to charge and discharge an arbitrary amount of electrolyte. Hence, in a flow battery, the charging and discharging power is defined by the power and the number of the individual cells, which are usually assembled to stacks. The deployed amount of electrolyte defines the energy storage capacity. Thus, the two most important characteristics of any battery, its nominal power and discharge capacity can be chosen independently from each other. This is the unique feature of a flow battery.

1.2.2 Advantages and disadvantages

The arbitrary ratio between power and storage capacity allows for an excellent adaptability of a flow battery to the intended use case. It is possible to change the ratio during the lifetime of the battery system by adding more cells or more electrolyte.

Also, the energy-related costs are low. For a large discharge capacity, the marginal costs of the Vanadium Redox Flow Battery (VRFB) tend to the expenses of the energy storage medium. According to [8], the material costs for the electrolyte of a VRFB is 2.47 $€L^{-1}$. Referred to the theoretical energy density of the standard vanadium electrolyte, which is 30 WhL^{-1}, this corresponds to specific energy-related costs of only 52 $€(kWh)^{-1}$. Today, the costs are indeed higher, mainly because significant production costs come on top of the electrolyte's material cost. Also, the usable specific energy density for a reasonable power density is substantially lower than the theoretical value. Hence, today, the costs for the vanadium electrolyte are in the range of 270 $€(kWh)^{-1}$ referred to the useable capacity [8]. To compare this cost on a system level, we also need to take the costs of the tanks into account. However, compared to the cost differences of lithium-based batteries on cell and system level, a flow battery has significant less overhead costs. In particular, it does not require cell balancing, single-cell monitoring, and cell packaging.

For mid-term to long-term storage, it is noteworthy that electrolyte, stored in the tanks, will not suffer from any self-discharge. Note, that this is not valid for the energy stored in the stacks themselves. However, in a reasonably dimensioned flow battery, this is only a small fraction of the totally stored energy.

Today, the high power related costs can be considered to be the biggest disadvantage of a VRFB. Due to the small number of units, assembling of stacks is manual labor.

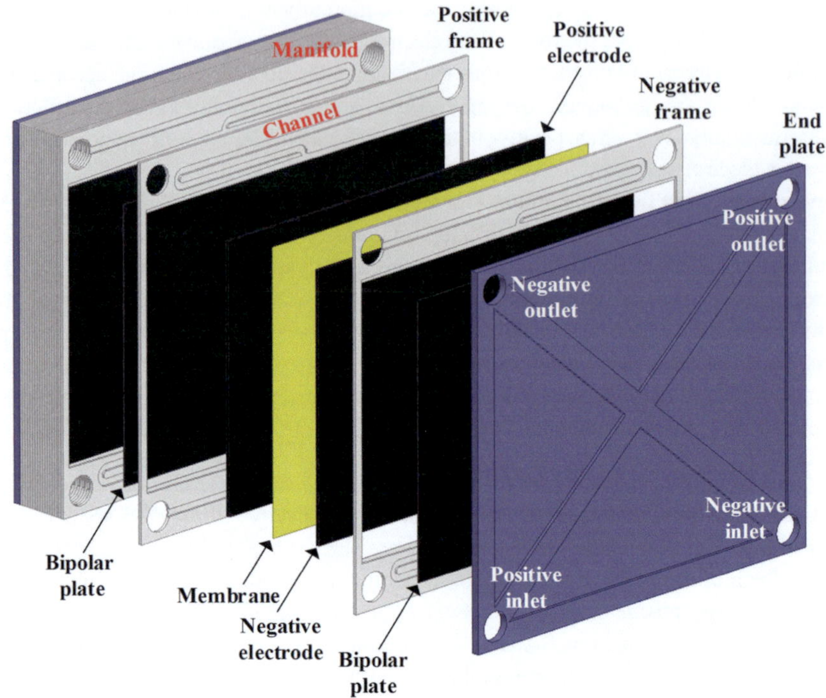

Figure 1-2: Structure and elements of generic flow battery cells, assembled to a stack

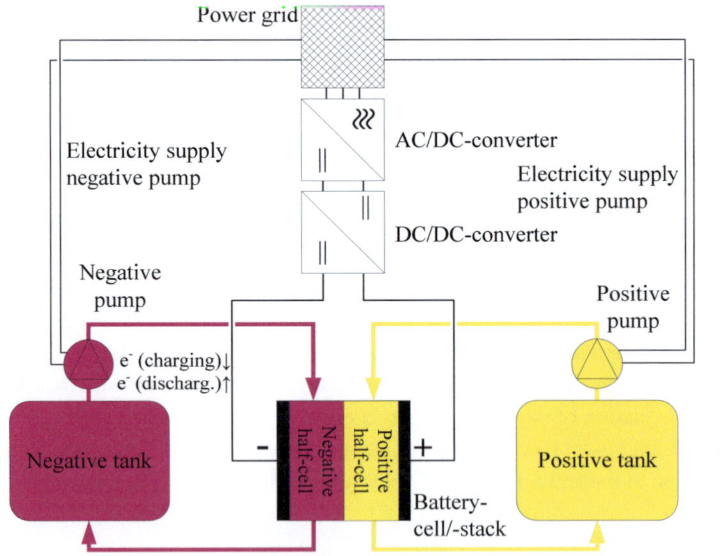

Figure 1-3: Composition of a flow battery system

Also, most companies only buy small quantities of the required materials from their suppliers. Hence, the stack-related costs are rather high. In [8], stack specific costs of more than 8,000 €(kW)$^{-1}$ are calculated. The costs for the pumps, the piping and the power conditioning system (PCS) further increase the power-related costs.

The PCS for a flow battery is usually more complex than for other types of batteries. Reasons are the low cell voltage, the limitation in the series connection of the cells and large voltage variations with state-of-charge (SoC) and charging/discharging currents. Finally, flow batteries suffer from a lower efficiency than lithium-based batteries. A comparatively high internal resistance, which also limits the power density, and additional loss mechanisms, such as the pump power demand, the crossover of vanadium ions across the membrane, as well as the shunt currents, reduce the VRFB efficiency.

1.2.3 Composition of a flow cell

In a flow cell, the electrode should provide a large contact area with the electrolyte. Therefore, porous graphite felts are predominantly used as electrodes. Between both electrodes, the ion exchange membrane is placed. As a single flow cell usually provides a low voltage of less than 2 V, a larger number of cells is connected in series electrically, as shown in Figure 1-2. This is efficiently realized by using so-called bipolar plates, made of a composite graphite material. These plates provide the planar electric connection of the electrode and simultaneously seal each positive half-cell from the successive negative half-cell. On top of the first and last bipolar plates, current collectors made from copper are placed. In the conventional stack design, massive aluminum endplates are used combination with threaded bolts, nuts and strong springs to clamp the stack. This is to obtain a high compact pressure all over the flow cells. The compact pressure reduces the contact resistances and improves the sealing. Finally, to electrically isolate the endplates from the stack, a plastic layer is placed between the current collector and the endplate. Note that the current collectors, the isolating plastic layer, as well as gaskets and bolts, used for the clamping, are not displayed in Figure 1-2.

The so-called frame, usually made from plastics using injection molding, provides the outer support. The frame also includes the inlet and outlet channels, which are required to partly decouple the electrolyte supply paths of the flow cells from each other.

Internally, the positive and the negative electrolytes are supplied and re-collected via four manifolds, commonly used by all flow cells of the stack.

1.2.4 Components of a flow battery system

A flow battery comprises two main units. The electrochemical active energy conversion unit, which is an arbitrary number of flow cells, and at least one pair of tanks, as shown in Figure 1-3. Figure 1-4 shows a universal flow battery design, in which six 30-cell stacks are connected to one pair of tanks with a volume of 8,000 L each. The system, rated 54 kW/216 kWh, is used in a model-based study [9].

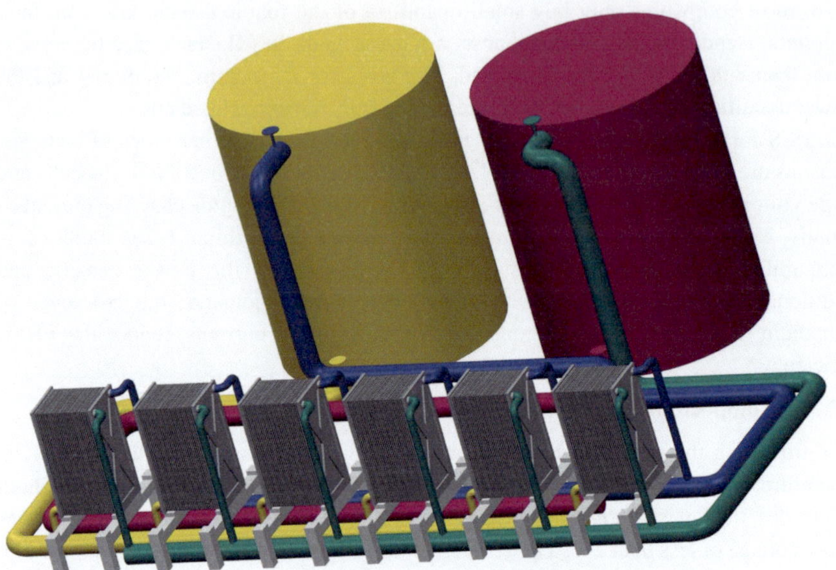

Figure 1-4: CAD scheme of a generic flow battery system with six stacks and two tanks

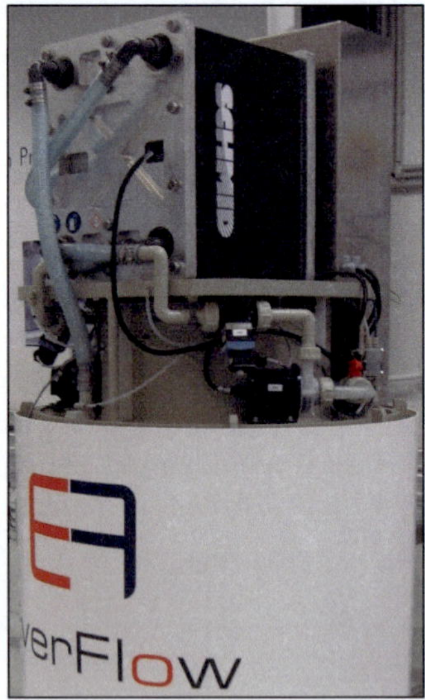

Figure 1-5: 2-kW/10-kWh VRFB prototype by SCHMID Energy Systems GmbH, Freudenstadt, Germany [10]

Figure 1-6: Gildemeister CellCube stacks;
Copyright by Gildemeister Energy Storage GmbH, used with permission

Figure 1-7: Interior view of a Gildemeister CellCube container;
Copyright by Gildemeister Energy Storage GmbH, used with permission

If the use case requires a more compact system, the stack is often placed on top of the tanks, as shown in Figure 1-5. Large systems consist of a large number of stacks and are often placed into containers, as shown in Figure 1-6 and Figure 1-7.

Additional components are usually two pumps, electric and hydraulic networks, as well as a PCS. A PCS normally consists of a DC/DC-converter (DC – direct current) and an AC/DC-converter (AC – alternating current). The DC/DC-converter adapts the low, load- and SoC-depending battery DC voltage to the high and constant DC voltage of the AC/DC-converter. The latter provides the conversion of the DC voltage into the AC voltage, existing in the power grid. The pumps can be supplied directly from the battery or from the power grid. Note, that the former is difficult if the battery is not used for a longer period and the energy stored in the stack is lost due to self-discharge.

1.2.5 A short history of flow batteries

In 1949, Dr. Walter Kangro patented a process for storing electric energy using a conversion cell and a reservoir [11]. In [12], he evaluates titanium and iron chloride as well as titanium and iron sulfate as two possible redox couples for the utilization in a flow battery.

In the early 1970s, the National Aeronautics and Space Administration (NASA) became interested in redox flow batteries [1]. They focused on the iron and chromium redox couple, covering many system-related topics, such as shunt currents and optimized flow rate control strategies (FRCS).

In the 1980s, Maria Skyllas-Kazacos pioneered the all-vanadium redox flow battery at the University of New South Wales (UNSW) in Australia. The corresponding patent was granted in 1988 [13]. One key finding was that up to 2 mol of pentavalent vanadium ions are dissolvable in 2 mol of sulfuric acid. This permits utilizing the transition metal vanadium, having four oxidation states, as a redox couple in both half-cells. In a VRFB, the fundamental redox reactions are as follows [14]. The fundamental processes are shown in Figure 1-8.

- Negative half-cell:

$$V^{2+} \underset{\text{charging}}{\overset{\text{discharging}}{\rightleftharpoons}} V^{3+}+e^{-}, E_{-}^{0} = -0.255 \text{ V} \tag{1-1}$$

- Positive half-cell:

$$VO_2^{+}+2H^{+}+e^{-} \underset{\text{charging}}{\overset{\text{discharging}}{\rightleftharpoons}} VO^{2+}+H_2O, E_{+}^{0} = 1.004 \text{ V} \tag{1-2}$$

- Total-cell reaction:

$$V^{2+}+VO_2^{+}+2H^{+} \underset{\text{charging}}{\overset{\text{discharging}}{\rightleftharpoons}} VO^{2+}+V^{3+}+H_2O, E^{0} = 1.259 \text{ V} \tag{1-3}$$

Using vanadium in both half-cells solved one key problem of previously studied redox couples, which was the crossover of ions over the membrane into the respective opposite half-cell. Although this process still occurs in the VRFB, it only leads to a certain self-discharge but does not damage or harm the electrolytes. Today, the VRFB is the most mature flow battery technology [15]. Reference [15] provides a very comprehensive presentation about the history of flow batteries and possible redox couples.

a) Charging process

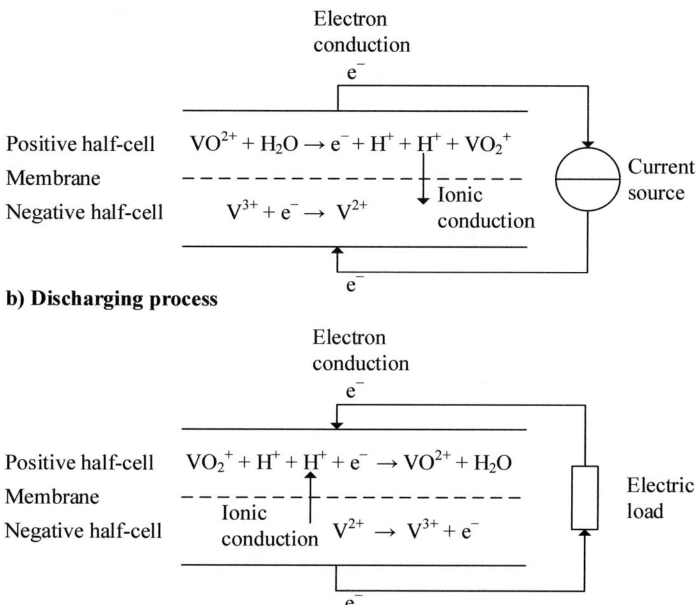

b) Discharging process

Figure 1-8: Fundamental redox reactions in a VRFB [14]

1.3   Objectives of the presented work

The presented work targets at increasing the VRFB system efficiency in both design and operation phase. Thereby, the reduction of losses caused by shunt currents, concentration overpotential and pump power demand is the overall goal. This goal is pursued using the following four steps:

- A multi-physical modeling approach for setting up a lumped-parameter model of a VRFB on a system level.
- A validation of the model with experimental data from VRFB manufacturers.
- An extensive holistic study on the impact of cell design on battery performance.
- An innovative flow rate control strategy.

1.4   Structure of the presented work

The presented work is organized as follows. In Chapter 2, a comprehensive multi-physical lumped-parameter model of the VRFB is presented. Each modeling aspect is illustrated with numerical examples to improve the understanding of the fundamental internal processes in a VRFB. Relevant definitions for the system design and operation are introduced in Chapter 3. Subsequently, the developed cell voltage model is extensively validated in Chapter 4 using measurement data from three different VRFB manufacturers.

In Chapter 5, twenty-four different cell designs are created using finite element analysis (FEA) simulations. The validated model and the derived cell designs are then deployed in a single-stack system design study in Chapter 6. In Chapter 7, the same is done for a system using a three-stack string. For both systems, a flow rate optimization is carried out simultaneously.

In Chapter 8, the author's previously published innovative flow rate control strategy [16] is summarized and a stack voltage controller is introduced for overcoming the drawback of a reduced discharge capacity. The improved method is extensively compared against the established flow rate control strategies. Finally, the work is summarized in Chapter 9 and an outlook is given, providing future research topics. Additional figures and calculations are included in the Appendix.

1.5 The flow factor

One key parameter used in this work is the so-called flow factor. Hence, it is shortly introduced at the very beginning. For operating a flow battery, a minimal flow rate is required. This flow rate is given by Faraday's first law of electrolysis. As shown in Section 8.2 on page 120, the minimal flow rate depends on the SoC as well as on sign and magnitude of the applied current. The flow factor scales the minimally required flow rate. Hence for a flow factor of five, the actually applied flow rate is five times larger than electrochemically required. Consequently, only 20 % (one fifth) of the particular vanadium ions partaking in the respective reactions are converted as the electrolyte flows through the cells. Thus, there is a large surplus of reactants, which facilitates the reactions.

Chapter 2

Flow battery modeling

2.1 Assumptions and simplifications

Modelling and simulating a full-scale VRFB system requires certain assumptions and simplifications. Each phenomenon is explained in more detail in the respective section that deals with the particular phenomenon.

- The temperature is constant.

- The flow rate is identical for both half-sides.

- Perfect mixture is obtained in tanks and cells.

- Oxygen and hydrogen evolution can be neglected.

- The volumes of the negative and the positive electrolyte remain constant.

- All cells of a stack are equally well supplied with electrolyte.

- The hydrogen proton concentration in both electrolytes is constant.

- Self-discharge reactions due to the diffusion of vanadium ions into the opposite half-cell are instantaneous.

- The mass transfer related to the crossover of vanadium ions across the membrane can be neglected for the short-term operation.

- The concentration of V^{2+} and VO_2^+ ions in the negative and positive half-cells, respectively, is always larger than zero.

- Both electrolytes have the same constant density and viscosity.

- The activation overpotential can be neglected.

2.2 State-of-charge of a Vanadium Redox Flow Battery

The state-of-charge (SoC) indicates how much electric charge the battery currently stores, referred to its theoretical capacity. For an ideal battery, the SoC is the integration of the injected or released electric current with respect to the time divided by the theoretical battery capacity, as shown in Eq. (2-1). An ideal battery does not lose any electric charge neither in the charging or the discharging process nor in times of standby. However, the theoretical capacity of a battery always differs from its nominal or useable capacity. This is mainly because the exploitation of the total SoC-range leads to several undesired effects such as accelerated aging or lower efficiencies. Thus, in regular operation, the SoC never reaches values of 0 % or 100 %.

$$SoC(t) = SoC(t_0) + \frac{1}{C_{\text{Theo}}} \int_{t_0}^{t_{\text{End}}} I(t)\,dt \qquad (2\text{-}1)$$

Wherein:

t	Time	(s)
t_0	Initial time	(s)
t_{End}	Time at end of calculations	(s)
SoC	State-of-charge	(-)
C_{Theo}	Theoretical capacity	(As)
I	Current	(A)

We can derive the theoretical capacity of a VRFB as shown in Eq. (2-2) [17]. The factor ½ is necessary because two electrolytes are required for a functioning flow battery. For the standard electrolyte with a total vanadium concentration of 1.6 molL^{-1}, we can derive a specific theoretical capacity of 21.4 AhL^{-1}. By multiplying this capacity with the open circuit voltage (OCV) of the electrolyte in a SoC of 50 %, which is around 1.4 V, we yield the capacity in WhL^{-1}. The standard electrolyte yields a specific capacity of 30 WhL^{-1}. However, we can only exploit this capacity if a SoC range between 0 % and 100 % is used with a completely lossless battery.

$$C_{\text{Theo}} = \frac{1}{2} V_{\text{Total}} c_{\text{V}} F \qquad (2\text{-}2)$$

Wherein:

C_{Theo}	Theoretical capacity	(As)
c_{V}	Total vanadium concentration	(molm^{-3})
F	Faraday constant	96,485 As·mol^{-1}
V_{Total}	Total electrolyte volume	(m^3)

In a VRFB, vanadium ions are reduced and oxidized during the charging and discharging process. As shown in the Eqs. (1-1) and (1-2), the completely discharged negative electrolyte with a SoC of 0 % only contains V^{3+} ions; the completely discharged positive electrolyte only contains VO^{2+} ions. Consequentially, the completely charged negative electrolyte with a SoC of 100 % only contains V^{2+} ions; the completely charged positive electrolyte only contains VO_2^+ ions. For a SoC between 0 % and 100 %, the negative and positive electrolyte consists of a fraction of the respective ion concentrations.

In this work, the internal electrochemical quantities are denoted with subscripts in the following manner. The first subscript indicates the particular vanadium ion. The subscript '2' denotes quantities referring to the vanadium ion V^{2+}. Consistently, the subscript '3' is used for the V^{3+} ions, '4' for the VO^{2+} ions and '5' for the VO_2^+ ions. The second subscript indicates the measurement location of the quantity. The subscript 'T' refers to the tank and the subscript 'C' refers to the cell. The third and final subscript indicates whether the quantity refers to the positive half-side (subscript '+'), or the negative half-side (subscript '−').

Because of the vanadium crossover, described in Section 2.4 on page 18, the SoCs of both electrolytes might deviate from each other during the battery operation. They are calculated as shown in the Eqs. (2-3) and (2-4) [18]. To facilitate the reading, from now on, the time variable is omitted.

$$SoC_{T-} = \frac{c_{2T-}}{c_{2T-} + c_{3T-}} \tag{2-3}$$

$$SoC_{T+} = \frac{c_{5T+}}{c_{4T+} + c_{5T+}} \tag{2-4}$$

The combined SoC of both tanks, SoC_T, can be expressed in two ways. Eq. (2-5) shows one way of a combined SoC. Herein, we just divide the concentration of the 'charged' species by the total concentration of vanadium ions in the two electrolytes. However, this SoC definition is not practical. If the V^{2+} ions are depleted in the negative tank, we can no longer operate the battery. Nevertheless, if the positive tank still stores VO_2^+ ions, the tank SoC according to Eq. (2-5) is not zero. Effectively, the positive tank still stores energy, but this energy is no longer accessible in this battery configuration.

$$SoC_T = \frac{c_{2T-} + c_{5T+}}{c_{2T-} + c_{3T-} + c_{4T+} + c_{5T+}} \tag{2-5}$$

The second way for deriving a combined SoC originates from the SoC measurement in practice. Using a bypass-cell which operates under open circuit conditions, we can constantly measure the open circuit voltage (OCV) of the two electrolytes [17]. VRFB manufacturers usually measure the OCV(SoC) correlation for their particular electrolyte/membrane combination. Using this data, it is possible to convert the measured OCV into the electrolyte SoC. Unfortunately, the OCV(SoC) relations supplied by the manufacturers are confidential. Hence, the Nernst equation, introduced in Section 2.6 on page 35, is used to describe the relationship between the OCV and the SoC. Under the assumption of an equal total vanadium concentration in both electrolytes, Eq. (2-6) can be derived. In the model, this relation is used to determine the combined tank SoC. For an extensive derivation of this correlation, the reader is referred to the Appendix, namely Section B.1 on page 151.

$$SoC_T = \frac{\sqrt{\dfrac{c_{2T-} \cdot c_{5T+}}{c_{3T-} \cdot c_{4T+}}}}{1 + \sqrt{\dfrac{c_{2T-} \cdot c_{5T+}}{c_{3T-} \cdot c_{4T+}}}} \tag{2-6}$$

2.3   Concentrations of the ionic species

The SoC calculation shows that the concentrations of the ionic species are the basic quantities in the VRFB. The species' concentration is affected by various processes. While the impact of the applied charging or discharging current on the concentration is intended, several undesired processes, so-called side reactions, also affect the concentrations. These side reactions include but are not limited to the oxidation of V^{2+} ions by atmospheric oxygen, the evolution of hydrogen and oxygen, the corrosion of carbon components, the precipitation of vanadium salts, the (electro) osmotic water drag and the vanadium crossover across the membrane as well as the shunt currents [19]. The effect of water drag and vanadium crossover can be reverted by mixing positive and negative tanks and re-conditioning the electrolytes. However, other side-reactions, such as the carbon corrosion or the precipitation of the vanadium salts are irreversible. Regarding parasitic processes that affect the species' concentration, only the phenomena of shunt currents and vanadium crossover are considered in this work. Finally, the coupling of the tanks and the cells via the volumetric flow rate also affects the concentrations of the ionic species in the tanks and the cells.

2.3.1 Tank concentration of vanadium ions

To calculate the tank concentration of the four different vanadium ions, we have to know the ionic flux, $J_{i\text{Stacks}}^{\text{out}}$, leaving the stacks and entering the tanks. Due to the shunt currents, the charging and discharging current of each cell is different, as laid out in Section 2.5 on page 26. It is assumed that both half-sides of all cells receive an identical flow rate. Hence, to yield the ionic flux of the total stack, we can summarize the concentrations of all cells of the stack and multiply them with the cell flow rate. The cell flow rate is the stack flow rate over the number of cells. The ionic flux of the total stack is shown in Eq. (2-7) for the negative half-side and in Eq. (2-8) for the positive half-side. To derive the total ionic flux of all stacks, the ionic fluxes of all stacks are summarized.

$$J_{i\text{Stacks}-}^{\text{out}} = \sum_{m=1}^{N_S} \frac{Q_{Sm}}{N_C} \sum_{n=1}^{N_C} c_{iCmn-}^{\text{out}} \tag{2-7}$$

$$J_{i\text{Stacks}+}^{\text{out}} = \sum_{m=1}^{N_S} \frac{Q_{Sm}}{N_C} \sum_{n=1}^{N_C} c_{iCmn+}^{\text{out}} \tag{2-8}$$

Wherein:

c_{iCmn}^{out}	Output concentration of vanadium species i in cell n of stack m	(molm^{-3})
i	Counting index for vanadium ions	(-)
$J_{i\text{Stacks}}^{\text{out}}$	Total ionic flux of vanadium species i from all stacks	(mols^{-1})
n	Counting index for cells	(-)
N_C	Number of cells per stack	(-)
N_S	Number of stacks per tank	(-)
m	Counting index for stacks	(-)
Q_{Sm}	Volumetric flow rate of stack m	(m^{3}s^{-1})

During the battery operation, we continuously take an ionic flux out of the tank and feed it to the stacks. Simultaneously, we collect the electrolyte coming out of the stacks and feed it back to the tank. Hence, the tank concentration of the different vanadium species i is determined by the outgoing and the incoming ionic fluxes, as shown in the Eqs. (2-9) and (2-10). The incoming ionic flux of the tank is coming out of the stacks and thus is denoted $J_{i\text{Stacks}}^{\text{out}}$. The outgoing ionic flux of the tank is the product of the tank volumetric flow rate and the tank concentration of the particular vanadium species. It is assumed that the ionic concentrations in the tank are homogenously distributed. Hence, the concentrations in the electrolyte leaving the tank with the flow rate Q_T are equal to the tank concentrations themselves.

$$V_T \frac{\mathrm{d}}{\mathrm{d}t} c_{iT-} = J_{i\text{Stacks}-}^{\text{out}} - Q_T c_{iT-}, \; i \in \{2,3,4,5\} \tag{2-9}$$

$$V_T \frac{\mathrm{d}}{\mathrm{d}t} c_{iT+} = J_{i\text{Stacks}+}^{\text{out}} - Q_T c_{iT+}, \; i \in \{2,3,4,5\} \tag{2-10}$$

Wherein:

c_{iT}	Concentration of vanadium species i in the tank	(molm^{-3})
J_i	Ionic flux of vanadium species i	(mols^{-1})
Q_T	Tank volumetric flow rate	(m^{3}s^{-1})
V_T	Tank volume	(m^3)

2.3.2 Tank concentration of hydrogen protons

Equation (1-2) indicates that besides vanadium ions, hydrogen protons also partake in the redox reactions. Hence, we have to consider their concentration as well. The oxidation of one VO^{2+} ion releases two H^+ protons, as shown in Eq. (1-2) on page 10. While one of them is required to cross the membrane to fulfill the balance of charges between the two half-cells, the second one remains in the positive half-cell [17]. Consequently, the concentration of hydrogen protons increases during the charging process in both half-cells.

The concentration of hydrogen protons in the positive and negative half-cell evolves as shown in Eq. (2-11) [17]. In this work, the Nafion 115 membrane is considered exclusively, which is a cation exchange membrane. Hence, the membrane is permeable for hydrogen protons resulting in identical proton concentrations on both half-sides.

The subscript 'H' is used for the quantities referring to hydrogen protons. The quantity $c_{HT+}(SoC{=}0)$ relates to the tank concentration of hydrogen protons in the completely discharged positive electrolyte, where no VO_2^+ ions are present.

$$c_{HT-} = c_{HT+} = c_{HT+}(SoC{=}0) + c_{5T+} \tag{2-11}$$

2.3.3 Cell concentration of vanadium ions considering vanadium crossover

In the flow cell, there are three different concentrations for every type of vanadium ion, namely the input concentration, the output concentration and the internal concentration. The input concentration corresponds to the concentration of the particular vanadium ion in the tank, c_{iT}. The output concentration, c_{iC}^{out}, is determined by the electric cell current, the vanadium crossover and the hydraulic coupling to the tank, as shown in Eq. (2-12) for the negative half-cell and in Eq. (2-13) for the positive half-cell. The resulting differential equations are solved using numerical integration. For deriving the initial values, the whole system is assumed to be in an equilibrium state and the electrolyte in cells and tanks is in a given start SoC.

The internal concentrations, c_{iC}, are derived using the arithmetic mean value of the particular tank concentration, c_{iT}, and the cell output concentration, c_{iC}^{out}. This corresponds to the assumption of a perfectly mixed cell [17].

$$V_E \frac{d}{dt} \begin{pmatrix} c_{2C-}^{out} \\ c_{3C-}^{out} \\ c_{4C-}^{out} \\ c_{5C-}^{out} \end{pmatrix} = \frac{1}{F} \begin{pmatrix} +I_C \\ -I_C \\ 0 \\ 0 \end{pmatrix} + \begin{pmatrix} J_{2Cross-} \\ J_{3Cross-} \\ J_{4Cross-} \\ J_{5Cross-} \end{pmatrix} + \frac{Q_S}{N_C} \begin{pmatrix} c_{2T-} - c_{2C-}^{out} \\ c_{3T-} - c_{3C-}^{out} \\ c_{4T-} - c_{4C-}^{out} \\ c_{5T-} - c_{5C-}^{out} \end{pmatrix} \qquad (2\text{-}12)$$

$$V_E \frac{d}{dt} \begin{pmatrix} c_{2C+}^{out} \\ c_{3C+}^{out} \\ c_{4C+}^{out} \\ c_{5C+}^{out} \end{pmatrix} = \frac{1}{F} \begin{pmatrix} 0 \\ 0 \\ -I_C \\ +I_C \end{pmatrix} + \begin{pmatrix} J_{2Cross+} \\ J_{3Cross+} \\ J_{4Cross+} \\ J_{5Cross+} \end{pmatrix} + \frac{Q_S}{N_C} \begin{pmatrix} c_{2T+} - c_{2C+}^{out} \\ c_{3T+} - c_{3C+}^{out} \\ c_{4T+} - c_{4C+}^{out} \\ c_{5T+} - c_{5C+}^{out} \end{pmatrix} \qquad (2\text{-}13)$$

Wherein:

c_{iC+}^{out}	Cell output concentration of Vanadium species i	(molm^{-3})
I_C	Cell current	(A)
J_{iCross}	Ionic flux of vanadium species i due to vanadium crossover and side reactions	(mols^{-1})
Q_S	Stack volumetric flow rate	(m^{3}s^{-1})
V_E	Electrode volume (Width × Height × Thickness × Porosity)	(m^3)

Figure 2-1 illustrates the three different concentrations for a sample cycle between a tank SoC of 20 % and 80 % and a current of 200 A. The electrolytes are assumed to contain a total vanadium concentration, c_V, of 1,600 molm^{-3}. During the charging process, the cell output concentration of V^{2+} ions rises compared to the tank concentration. This is because of the oxidation of V^{2+} ions to V^{3+} ions by the charging current. Consistently, this elevates the internal cell concentration as well. During the discharging process, the internal and the output concentration of V^{2+} ions is lower than the V^{2+}-concentration in the tank. Finally, it is noticeable that the three different concentrations differ more strongly for a lower flow rate than for a higher flow rate.

2.4 Vanadium crossover

A membrane separates the positive and the negative half-cell. Its main task is to prevent the negative and positive electrolytes from mixing. However, the exchange of certain ions is required for the balance of charges. The ability of letting certain ions pass while blocking the others is called selectivity.

Unfortunately, membranes do not have a perfect selectivity. Driven by a pressure gradient (convection), an electric field (migration) or a concentration gradient (diffusion), electrolyte or certain parts of the electrolyte may cross the membrane [20].

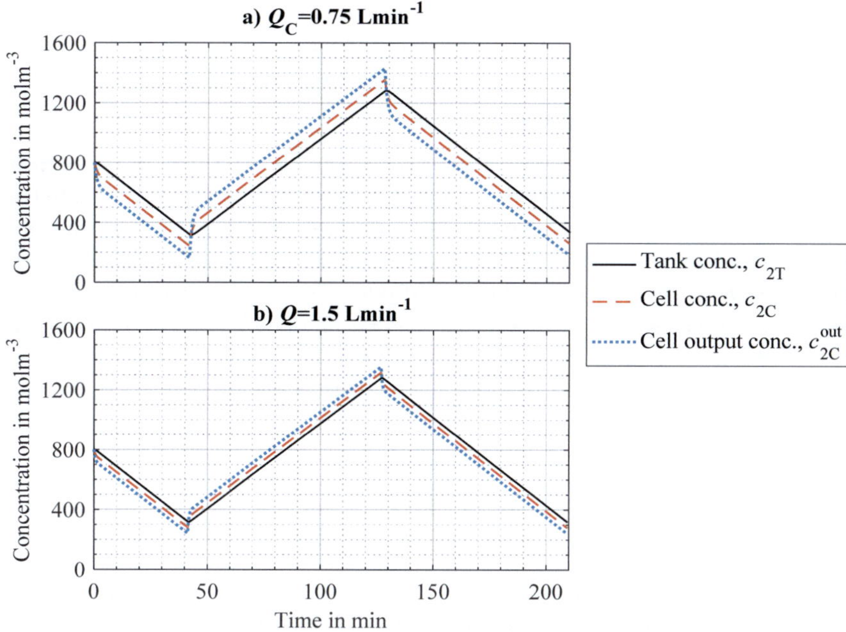

Figure 2-1: V^{2+} concentrations during a sample cycle with 200 A between tank SoC 20 % and 80 % for two different constant flow rates

In general, vanadium crossover and water transfer through the membrane is modeled using the Nernst-Planck and the Schlogl equation, which accounts for all three of the mechanisms mentioned before [18, 21–23].

For a Nafion 117 membrane, it is observed that the diffusion phenomenon causes 90-95 % of the total vanadium crossover [18, 24]. Thereby the effect of migration is found to play a minor role. In practice, the viscosities of positive and negative electrolyte deviate from each other. Hence, an identical flow rate for positive and negative half-cell introduces a pressure gradient between two half-cells. This gives way to the crossover caused by convection. Different flow rates for the two half-cells, reflecting the different viscosities, eliminate the effect of convection [25].

For studies on a stack level, the approaches mentioned above are too complex to be solved with reasonable computational effort in a reasonable amount of time. Hence, several simplified methods have been developed. In [24], the Nernst-Planck equation and Schlogl's equation are solved for a zero-dimensional model. This model accounts for convection, migration and diffusion of vanadium ions through the membrane. However, the comparison with the formerly presented two-dimensional model [18] reveals large differences between both modeling approaches. A similar approach is employed in [26], but the results are not valid as the work contains a unit conversion error and thus the results are off by a factor of 1,000.

Many other system level studies only consider the effect of diffusion, which is modeled using Fick's law [16, 19, 27–31]. The presented models assume that no actual mass transfer across the membrane takes place which is also assumed in this work.

2.4.1 Vanadium crossover flux

Although the total vanadium concentration is the same in the negative and positive electrolytes, a concentration gradient exists because the vanadium ions exist in different oxidation states on the positive and negative half-sides. Driven by this concentration gradient, vanadium ions may cross the membrane and diffuse into the opposite half-cell. This diffusion can be described using the diffusion flux density, j_i, of the vanadium species i, described by Fick's first law, shown in Eq. (2-14) [28, 32]. The diffusion flux density depends on the gradient of the concentration c_i and the diffusion coefficient D_i. The diffusion coefficient is a measure for the membrane's resistivity against the diffusion of vanadium ions.

$$j_i = D_{i\mathrm{Mem}} \frac{dc_i}{dx}, \; i \in \{2,3,4,5\} \tag{2-14}$$

Wherein:

c_i	Concentration of vanadium species i	(molm^{-3})
$D_{i\mathrm{Mem}}$	Diffusion coefficient of vanadium species i in the membrane	(m^{2}s^{-1})
j_i	Ionic flux density of vanadium species i through the membrane	(mols^{-1}m^{-2})
x	Location	(m)

In this work, no spatial distribution of any process is considered. Within each half-cell, perfect mixture is assumed to take place. Thus, the concentration gradient is the difference in the respective species' concentration on both half-sides, divided by the membrane thickness, δ_{Mem}. Consequently, we calculate the effective ionic flux, J_{iC}, by multiplying the flux density with the membrane area, A_{Mem}. The membrane area corresponds to the electrode area, A_{E}, as shown in Eq. (2-15) for the negative half-cell and in Eq. (2-16) for the positive half-cell.

$$J_{iC-} = D_{i\mathrm{Mem}} \frac{c_{iC+} - c_{iC-}}{\delta_{\mathrm{Mem}}} A_{\mathrm{Mem}}, \; i \in \{2,3,4,5\} \tag{2-15}$$

$$J_{iC+} = D_{i\mathrm{Mem}} \frac{c_{iC-} - c_{iC+}}{\delta_{\mathrm{Mem}}} A_{\mathrm{Mem}}, \; i \in \{2,3,4,5\} \tag{2-16}$$

2.4.2 Self-discharging reactions

If vanadium ions enter the respective opposite half-cell, they partake in redox reactions as long as the required reaction partners are present [32, 33]. We can express these reactions as follows.

- Crossover of VO^{2+} or VO_2^+ ions from the positive into the negative half-cell:

$$VO_2^+ + V^{3+} \rightarrow 2VO^{2+} \tag{2-17}$$

$$VO_2^+ + 2V^{2+} + 4H^+ \rightarrow 3V^{3+} + 2H_2O \tag{2-18}$$

$$VO^{2+} + V^{2+} + 2H^+ \rightarrow 2V^{3+} + H_2O \tag{2-19}$$

- Crossover of V^{2+} or V^{3+} ions from the negative into the positive half-cell:

$$V^{2+} + VO^{2+} + 2H^+ \rightarrow 2V^{3+} + H_2O \tag{2-20}$$

$$V^{2+} + 2VO_2^+ + 2H^+ \rightarrow 3VO^{2+} + H_2O \tag{2-21}$$

$$V^{3+} + VO_2^+ \rightarrow 2VO^{2+} \tag{2-22}$$

2.4.3 Ionic flux resulting from vanadium crossover and self-discharging reactions in the negative half-cell

For each vanadium species, we have to consider the vanadium ions which leave or enter the respective half-cell due to the vanadium crossover. The resulting redox reactions oxidize and reduce vanadium ions, which we can express as an additional ionic flux into or out of the respective half-cell as well.

In reality, all reactions occur simultaneously and not in a specific order. However, to compute the resulting reaction fluxes in one single step, we have to put the reactions into a certain order.

In the negative half-cell, reaction (2-17) yields VO^{2+} ions as one of the reaction products. In normal operation, the battery SoC does not reach a value of 0 %, meaning that there are always sufficient V^{2+} ions present to let reaction (2-19) take place. Hence, the VO^{2+} ions resulting from reaction (2-17) further react to V^{3+} ions. Therefore, when put into an order, reaction (2-17) should not be considered at last, because then VO^{2+} ions are present in the negative half-cell at the end of the calculation time step.

There are two reasonable orders for the sequence of the reactions, namely (2-17)→(2-18)→(2-19) and (2-18)→(2-17)→(2-19). In the first order, reaction (2-18) will not take place, since we can assume that there are always enough V^{3+} ions present in the negative half-cell to reduce all present VO_2^+ ions to VO^{2+} ions (reaction (2-17)). Thus, reaction (2-18) lacks the reaction partner VO_2^+ and the self-discharge process is reduced to the order (2-17)→(2-19).

In normal operation, the second order can also be simplified. If the negative electrolyte is not completely discharged, there will be sufficient V^{2+} ions to reduce all entered VO_2^+ ions to V^{3+} ions. In this case, reaction (2-17) is skipped because no more VO_2^+ ions are present to react with V^{3+} ions. Hence, the reactions are considered in the order (2-18)→(2-19).

The order in which the self-discharge reactions are considered is irrelevant for the resulting net ionic fluxes, as shown in Figure 2-2. In the first order, *a* VO^{2+} ions and

b VO_2^+ ions diffuse into the negative half-cell and react to $(4b + 2a)$ V^{3+} ions. Thereby, b V^{3+} and $(a+2b)$ V^{2+} partake in the reactions. As b V^{3+} ions are required in the reactions, the VO^{2+} and VO_2^+ ions diffusing into the negative half-cell effectively react to $(3b + 2a)$ V^{3+} ions in total. This process is depicted in Figure 2-2 a).

The second order represents the simpler one, as shown in Figure 2-2 b). The VO^{2+} and VO_2^+ ions diffusing into the negative half-cell react to $(3b + 2a)$ V^{3+} ions in total, which is equivalent to the first order. Because of its simpler sequence, the second order is implemented into the model.

a) Negative half-cell, reaction order (2-17) $\rightarrow$ (2-18) (skipped) $\rightarrow$ (2-19)

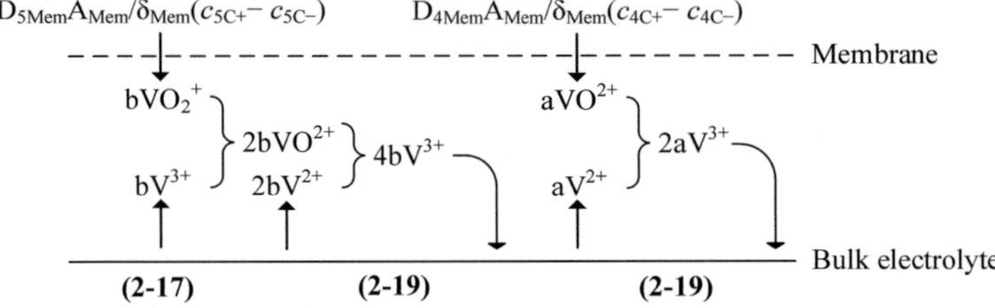

b) Negative half-cell, reaction order (2-18) $\rightarrow$ (2-17) (skipped) $\rightarrow$ (2-19)

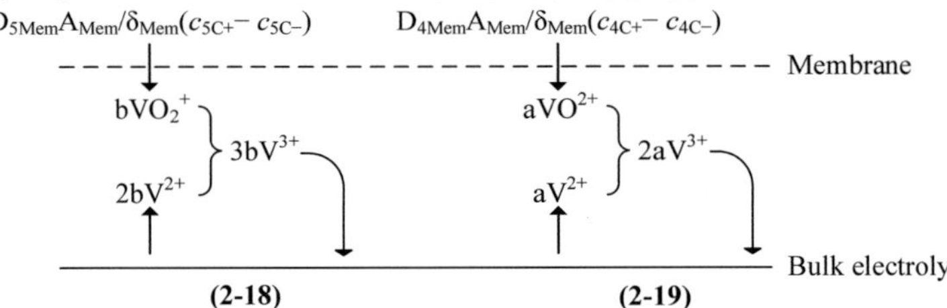

Figure 2-2: Self-discharge reactions due to the vanadium crossover in the negative half-cell

V^{2+} ions are forced to leave the negative half-cell, because in the positive half-cell, no ions in this oxidation state are present and thus a concentration gradient exists. We can calculate the resulting ionic flux using Eq. (2-15), yielding Eq. (2-23).

$$J_{2C-} = D_{2\mathrm{Mem}} \frac{c_{2C+} - c_{2C-}}{\delta_{\mathrm{Mem}}} A_{\mathrm{Mem}} \tag{2-23}$$

Now we are interested in the flux of VO^{2+} and VO_2^+ ions penetrating the negative half-cell through the membrane from the positive half-cell. These fluxes are given by the Eqs. (2-24) and (2-25).

$$J_{4C-} = D_{4\mathrm{Mem}} \frac{c_{4C+} - c_{4C-}}{\delta_{\mathrm{Mem}}} A_{\mathrm{Mem}} \tag{2-24}$$

$$J_{5C-} = D_{5\mathrm{Mem}} \frac{c_{5C+} - c_{5C-}}{\delta_{\mathrm{Mem}}} A_{\mathrm{Mem}} \tag{2-25}$$

If we take a look at reaction (2-18), we see that each VO_2^+ ion reacts with two V^{2+} ions to three V^{3+} ions. Thereby, all entered VO_2^+ ions are converted. This means that the effective ionic flux of VO_2^+ ions, $J_{5Cross-}$, is zero. After all calculations are carried out for a particular time step, VO_2^+ ions are no longer present in the negative half-cell. Simultaneously, a negative flux arises for the V^{2+} ions and a positive flux arises for the V^{3+} ions. The ionic flux of the V^{2+} ions is caused by reaction (2-18) and is $-2(c_{5C+} - c_{5C-})D_{5Mem}$. Thus, it is twice the ionic flux of the VO_2^+ ions penetrating the negative half-cell due to the vanadium crossover. The ionic flux of the V^{3+} ions is three times the ionic flux of the VO_2^+ ions.

Now, we take a look at reaction (2-19). Each present VO^{2+} ion reacts with one V^{2+} ion to two V^{3+} ions. Hence, the VO^{2+} ions penetrating the negative half-cell through the membrane are now converted into V^{3+} ions. Thus, the effective ionic flux of VO^{2+} ions, $J_{4Cross-}$, is zero as well. The number of V^{2+} ions is reduced by the number of VO^{2+} ions. Hence, the ionic flux of V^{2+} ions due to the reaction (2-19) is $-(c_{4C+} - c_{4C-})D_{4Mem}$. The ionic flux of V^{3+} ions is positive with twice the magnitude. Effectively, the ionic flux of V^{2+} ions because of the self-discharge reactions is $-(c_{4C+} - c_{4C-})D_{4Mem} - 2(c_{5C+} - c_{5C-})D_{5Mem}$.

The number of V^{3+} ions in the negative half-cell is increased by the self-discharge reactions. The entered VO_2^+ ions each react to three V^{3+} ions in the reaction (2-18). Further, each entered VO^{2+} ion reacts to two additional V^{3+} ions due to reaction (2-19). In total, the V^{3+} flux due to the self-discharge reactions is $2(c_{4C+} - c_{4C-})D_{4Mem} + 3(c_{5C+} - c_{5C-})D_{5Mem}$.

Together with the fluxes of V^{2+} and V^{3+} ions leaving the negative half-cell and diffusing into the positive half-cell, the fluxes can be put into a matrix form, as shown in Eq. (2-26), which allows for an efficient implementation in MATLAB.

$$
\begin{pmatrix} J_{2Cross-} \\ J_{3Cross-} \\ J_{4Cross-} \\ J_{5Cross-} \end{pmatrix} =
$$

$$
\frac{A_{Mem}}{\delta_{Mem}} \begin{pmatrix} (c_{2C+} - c_{2C-})D_{2Mem} - (c_{4C+} - c_{4C-})D_{4Mem} - 2(c_{5C+} - c_{5C-})D_{5Mem} \\ (c_{3C+} - c_{3C-})D_{3Mem} + 2(c_{4C+} - c_{4C-})D_{4Mem} + 3(c_{5C+} - c_{5C-})D_{5Mem} \\ 0 \\ 0 \end{pmatrix} \tag{2-26}
$$

2.4.4 Ionic flux resulting from vanadium crossover and the self-discharging reactions in the positive half-cell

The aforementioned considerations are conducted in an analogue manner for the positive half-cell. The self-discharge reaction (2-20) yields V^{3+} ions, which further react with VO^{2+} ions in the self-discharge reaction (2-22). Hence, reaction (2-20) should not be considered at last. Consequently, there are two reasonable orders of the reactions, namely (2-20)→(2-21)→(2-22) and (2-21)→(2-20)→(2-22).

In the first order, the reaction (2-20) converts all V^{2+} ions diffusing into the positive half-cell in a particular time step into V^{3+} ions. Hence, V^{2+} ions are no longer present in the positive half-cell and reaction (2-21) is skipped. Consequently, the self-discharge process is reduced to the order (2-20)→(2-22).

In the second order, reaction (2-21) converts all V^{2+} ions, existing in the positive half-cell, into VO^{2+} ions. Thus, reaction (2-20) is skipped and the reaction order is simplified to (2-21)→(2-22).

Again, the order in which we consider the reactions to take place is irrelevant for the arising net ionic fluxes, as shown in Figure 2-3. In the first case, a V^{2+} and b V^{3+} ions diffuse into the positive half-cell in a particular time step. In total, a VO^{2+} and $(2a+b)$ VO_2^+ ions react with these ions to $(4a+2b)$ VO^{2+} ions, as shown in Figure 2-3 a). As a VO^{2+} ions are required for the self-discharge reactions, the net change in the number of VO^{2+} ions is only $(3a+2b)$.

The second order is the simpler one, as shown in Figure 2-3 b). The a V^{2+} and b V^{3+} ions diffusing into the positive half-cell in the considered time step react with $(2a+b)$ VO_2^+ ions to $(3a+2b)$ VO^{2+} ions. Hence, the net change in the number of ions is equivalent to the one of the first order. Because of its simpler sequence, the second order is implemented into the model.

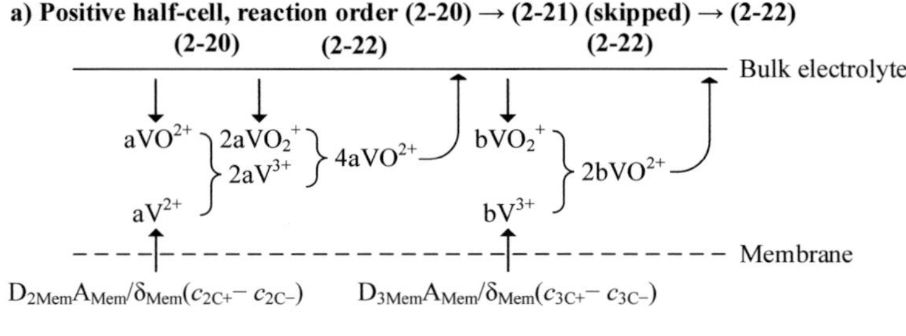

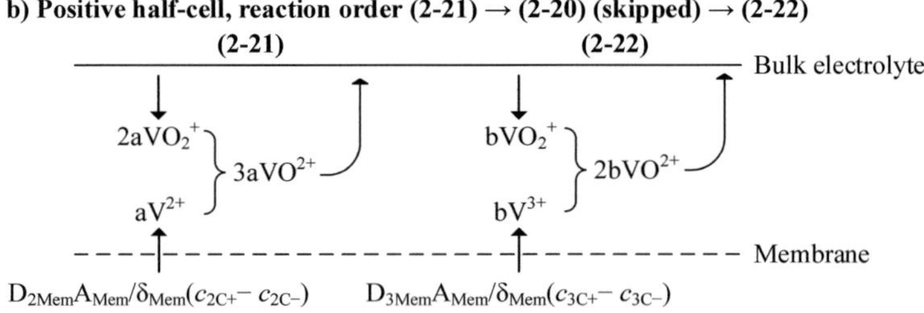

Figure 2-3: Self-discharge reactions due to the vanadium crossover in the positive half-cell

In the positive half-cell, all entered V^{2+} and V^{3+} ions are immediately converted into VO^{2+} ions. Thus, during the normal operation, no V^{2+} and V^{3+} ions are present in the positive half-cell. The effective ionic fluxes are derived in an analogue manner as for the negative half-cell, as shown in Eq. (2-27).

$$\begin{pmatrix} J_{2\text{Cross}+} \\ J_{3\text{Cross}+} \\ J_{4\text{Cross}+} \\ J_{5\text{Cross}+} \end{pmatrix} =$$

$$\frac{A_{\text{Mem}}}{\delta_{\text{Mem}}} \begin{pmatrix} 0 \\ 0 \\ 3(c_{2\text{C}-} - c_{2\text{C}+})D_{2\text{Mem}} + 2(c_{3\text{C}-} - c_{3\text{C}+})D_{3\text{Mem}} + (c_{4\text{C}-} - c_{4\text{C}+})D_{4\text{Mem}} \\ -2(c_{2\text{C}-} - c_{2\text{C}+})D_{2\text{Mem}} - (c_{3\text{C}-} - c_{3\text{C}+})D_{3\text{Mem}} + (c_{5\text{C}-} - c_{5\text{C}+})D_{5\text{Mem}} \end{pmatrix}$$

(2-27)

2.4.5 Numerical example

In short-term, the vanadium crossover influences the battery operation only in terms of efficiency. The vanadium crossover and the self-discharge reactions which it triggers in the opposite half-cell result in an equivalent self-discharging current, as shown in Figure 2-4.

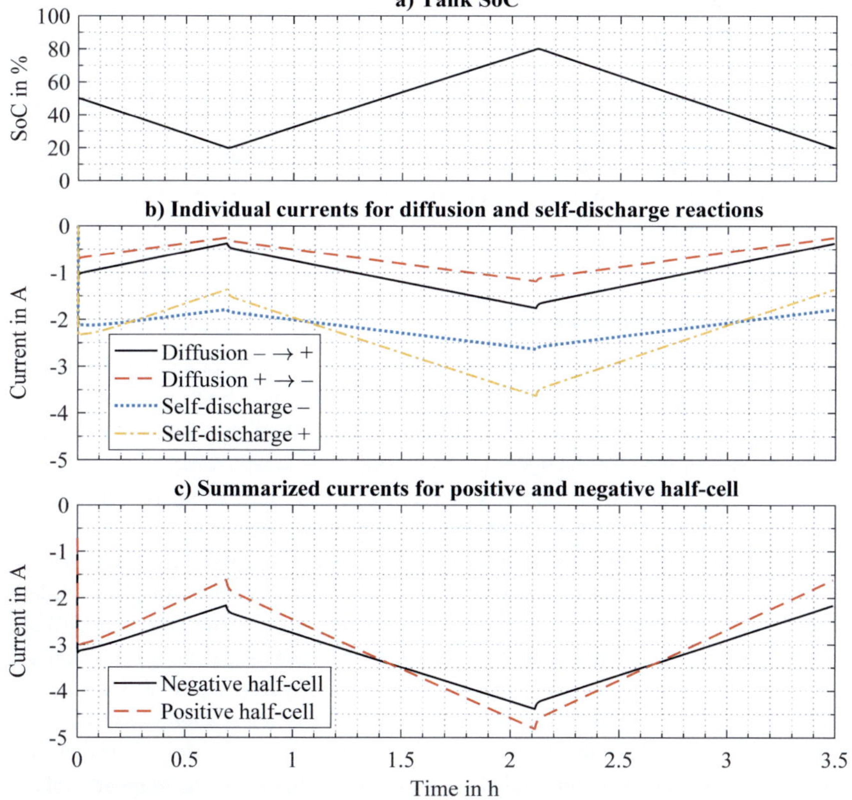

Figure 2-4: Equivalent currents due to the vanadium crossover through the membrane of a single 2000-cm^2 cell and respective self-discharge reactions in the opposite half-cell

Therein, the simulation results of a charging/discharging cycle between a tank SoC of 20 % and 80 % are shown. For the cycle simulation, a tank volume of 10 L per half-side, a total vanadium concentration of 1.6 $molL^{-1}$, a constant flow rate of 1.5 $Lmin^{-1}$ and a charging/discharging current of ±200 A is used. Additional parameters which remain unchanged throughout this work, including the diffusion coefficients of the simulated Nafion 115 membrane, are given in the Appendix B.2 on page 152.

In Figure 2-4, the ionic fluxes due to the vanadium crossover are interpreted as equivalent electric currents. The magnitude of the currents strongly depends on the SoC. A higher SoC increases the crossover losses. This is reasonable, as a higher SoC corresponds to a higher concentration of V^{2+} and VO_2^+ ions. Thus, more V^{2+} and VO_2^+ ions are going to cross the membrane. In the half-cell from which they originate, this is equivalent to a discharging process. Also, they trigger self-discharge reactions in the half-cell in which they arrive, which corresponds to a discharging process there as well. From the current magnitude, we see that the self-discharge processes result in a higher equivalent discharging current than the diffusion of the ions itself.

The crossover of V^{2+} and VO_2^+ ions is nearly zero when the SoC is low. However, at this point, the crossover of V^{3+} and VO^{2+} ions is dominant, because of their higher concentrations. In the half-cell in which they arrive, these ions also trigger self-discharge reactions.

As shown in Figure 2-4 c), the parasitic currents are different for the negative and the positive half-side. Hence, in long-term operation, the SoCs of the two half-sides might deviate from each other.

2.5 Shunt currents

2.5.1 Shunt current phenomenon

In order to obtain a reasonable battery voltage, we have to connect several cells electrically in series. Regarding the hydraulic interconnection, both series and parallel connections are possible in principle. However, in a hydraulic series connection, the total flow rate, required for the operation of all cells, has to pass every single cell. This leads to a very high pressure drop and thus to a very high pump power demand. Therefore, in general, the parallel connection is preferred [17].

The disadvantage of combining an electric series connection and a hydraulic parallel connection is the occurrence of shunt currents. Within a stack, mutual manifolds usually supply the cells with electrolyte. Hence, the electrolyte path connects cells with different electric potentials to each other. As the electrolyte is a relatively good ionic conductor, shunt currents evolve. Using the electrolyte supply path, hydrogen protons are now able to skip one or several cells. Thus, they do not partake in the desired electrochemical reaction in the skipped cells. The stored energy, namely the product of the electric charge of the hydrogen proton and the sum of the voltages of the skipped cells, is converted into heat and thus lost for the battery operation. Therefore, shunt currents lower the battery efficiency.

The shunt current phenomenon was first observed in electrochemical processes such as chloralkaline electrolysis [34]. For this and comparable processes from the field of electro synthesis, a series arrangement of several cells is common as well [35]. Beside energy loss, shunt currents have also be found to cause additional negative effects, such as material corrosion [36, 37].

2.5.2 Shunt current modeling

Within the stack, four manifolds, two per half-side, transport the electrolyte to the cells and collect it when it leaves the cells, as shown in Figure 2-5. The so-called channels connect each cell to its inlet and outlet manifolds. The channels are usually carried out in the shape of a meander. This intentionally extends the electrolyte path to increase the shunt current resistance. The disadvantage of the channels is an increased pressure drop within the cell and hence an increased pump power demand [38].

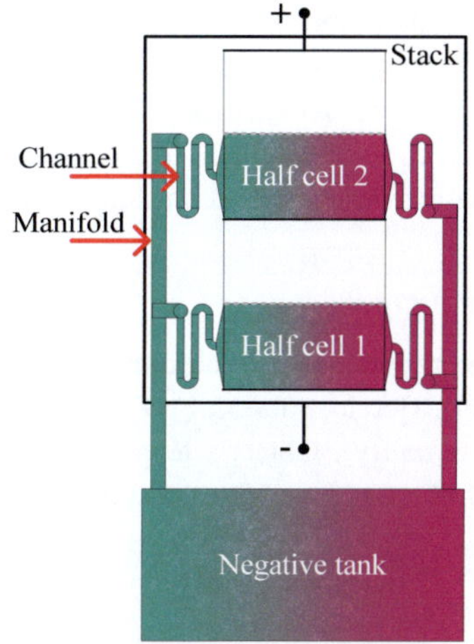

Figure 2-5: Schematic of the negative electrolyte supply of a two-cell stack

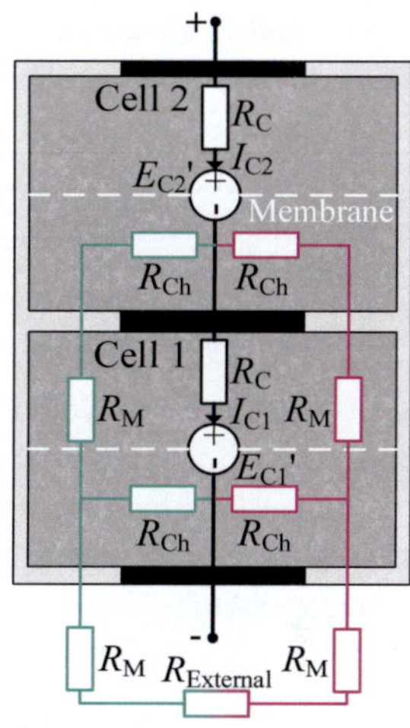

Figure 2-6: Equivalent electric circuit of the negative half-side in a two-cell stack

To compute the shunt currents, the equivalent electric circuit of the stack with its electric and hydraulic components is set up, as shown in Figure 2-6 [9, 31, 39, 40]. It contains the actual cell resistance (R_C), as well as the effective resistances of the manifolds (R_M), the channels (R_{Ch}) and the external hydraulic circuitry (tanks, pipes, etc. – $R_{External}$). While the cell resistance is assumed to be constant, the resistances of the manifolds and the channels vary with the SoC of the negative and the positive electrolyte.

2.5.3 Conductivity of negative and positive electrolyte

The conductivities of negative and positive electrolyte, σ_- and σ_+, differ strongly from each other. Further, they are temperature- and SoC-dependent [40, 41]. The conductivity of the standard electrolyte has been measured and published by two research groups. The Eqs. (2-28) to (2-31) describe the conductivity for a temperature of 298 K.

- Fraunhofer ICT, Germany [40]:

$$\sigma_- = 19.6 \text{ Sm}^{-1} + 10.7 \text{ Sm}^{-1} \cdot SoC_- \tag{2-28}$$

$$\sigma_+ = 30.8 \text{ Sm}^{-1} + 14.6 \text{ Sm}^{-1} \cdot SoC_+ \tag{2-29}$$

- University of New South Wales, Australia [41]:

$$\sigma_- = 18.8 \text{ Sm}^{-1} + \ 7.3 \text{ Sm}^{-1} \cdot SoC_- \tag{2-30}$$

$$\sigma_+ = 28.9 \text{ Sm}^{-1} + 13.9 \text{ Sm}^{-1} \cdot SoC_+ \tag{2-31}$$

In this work, the arithmetic mean value of both measurements is employed, as shown in the Eqs. (2-32) and (2-33):

$$\sigma_- = 19.2 \text{ Sm}^{-1} + \ 9.0 \text{ Sm}^{-1} \cdot SoC_- \tag{2-32}$$

$$\sigma_+ = 29.9 \text{ Sm}^{-1} + 14.3 \text{ Sm}^{-1} \cdot SoC_+ \tag{2-33}$$

Herein, SoC_- and SoC_+ denote the SoC of the negative and the positive electrolyte, respectively, in the particular object where the resistance is computed. Due to the strong impact of the SoC on the electrolyte conductivity, the model explicitly considers the different SoCs in the battery system. Hence, the electrolyte conductivity is calculated for each inlet and outlet channel, as well as for the forward and backward piping from and to the tanks.

2.5.4 Shunt current resistance of manifolds and pipes

The electric resistances of hydraulic elements are computed using their dimensions and the electrolyte conductivity, as shown in Eq. (2-34). Herein, subscript 'O' denotes the respective object (channel, manifold or external circuitry). Object's length is l_O, and object's cross sectional area is CSA_O. The quotient of length over cross sectional area is called geometry factor.

$$R_{O+/-} = {}^1\!/_{\sigma_{+/-}} \cdot \frac{l_O}{CSA_O} \tag{2-34}$$

2.5.5 Shunt current resistance of the channels

For the meander-shaped channel, the geometry factor is derived using finite-element-analysis (FEA). The cell, consisting of inlet channel, distribution funnel, porous flow-through graphite electrode, collection funnel and outlet channel is modeled using the ANSYS Workbench. To determine the geometry factor, the cell model is set up in ANSYS Maxwell, which allows for the computation of electric fields and currents. ANSYS Maxwell is in particular suited for this task because it supports a flexible geometry parametrization. Hence, it adapts the cell model to the key parameters electrode width, height, and thickness, manifold diameter, channel height and channel width.

For determining the geometry factor, the following procedure is developed. An ideal conductor represents the electrode to which an excitation voltage, $E_{Test} = 10$ V is applied. A test material with a conductivity of $\sigma_{Test} = 10\ \mathrm{Sm^{-1}}$ is assigned to funnel and channel of electrolyte inlet and outlet. The dimensions of the manifolds are taken into account, but the manifold itself is not part of the 3D-model. An excitation voltage of 0 V is applied to the faces which represent the beginning of the inlet channel and the end of the outlet channel. The described arrangement is shown in Figure 2-7.

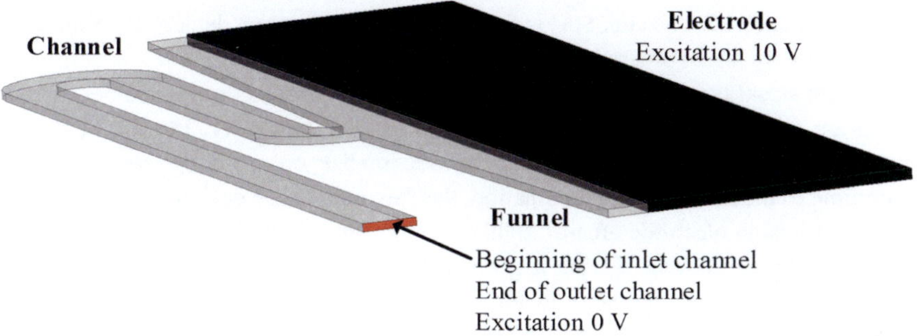

Figure 2-7: Excitation boundaries in the 3D-cell model for the calculation of the power loss

The chosen 'DC-Conduction' solver of ANSYS Maxwell evaluates the total losses, P_{Loss}, in the setup. The maximum permitted number of iterations is 100. A percentage error of 0.01 or smaller indicates the convergence of the solution.

By applying Eq. (2-35), the power loss, which is the result of the FEA, is converted into the required ratio of channel length over channel cross-sectional area, which is the channel geometry factor. The channel geometry factor also comprises the funnel. However, the funnel has a negligible impact on the total geometry factor because of its short and wide dimensions.

$$\frac{l_{Ch}}{CSA_{Ch}} = 2\frac{\sigma_{Test}E_{Test}^{2}}{P_{Loss}} = \frac{2000\,\mathrm{Wm^{-1}}}{P_{Loss}} \qquad (2\text{-}35)$$

2.5.6 Shunt current resistance of the manifold

The manifold is considered as a short and wide cylinder filled with electrolyte. Hence, its geometry factor can be computed analytically, as shown in Eq. (2-36). In this work, a constant total cell thickness of 10 mm is assumed. The manifold diameter changes according to the cell design, as laid out in Section 5.2.1 on page 78. The deployed values are given in Table 5-1 on page 78.

$$\frac{l_{M}}{CSA_{M}} = \frac{\delta_{C}}{\pi\left(\frac{d_{M}}{2}\right)^{2}} \qquad (2\text{-}36)$$

Wherein:
δ_{C} Total cell thickness   (m)
d_{M} Manifold diameter   (m)

2.5.7   Numerical example of shunt currents in a single stack

2.5.7.1 Equivalent shunt current

For all examples in this section, the VRFB is supplied with a flow rate corresponding to five times the stoichiometrically required flow rate for the particular operation. For more information on this flow rate control strategy, the reader is referred to Section 8.2, starting on page 120. The tank SoC is fixed at 50 % and a current density of 75 mAcm^{-2} is applied, if not stated otherwise.

In a single stack, the shunt current magnitude mainly depends on the number of cells which are electrically connected in series and the cell geometry. For the numerical examples, two geometries introduced in Section 5.3 on page 81 are employed. According to the introduced denomination, they are denoted as design 2.1 and 4.6. Cell design 2.1 has an electrode area of 2000 cm^2, one and a half channel meanders and a channel width of 20 mm. Having a geometry factor of 11,644 m^{-1}, it represents the design with the lowest geometry factor and thus with the largest shunt currents. Cell design 4.6 has an electrode area of 4000 cm^2, two and a half channel meanders and a channel width of 10 mm. Having a geometry factor of 52,159 m^{-1}, it represents the design with the highest geometry factor and thus the smallest shunt currents. Both designs are used to simulate a single stack with up to 40 cells, as shown in Figure 2-8.

To evaluate the designs regarding shunt currents, the equivalent shunt current is introduced as shown in Eq. (2-37). We can directly relate the equivalent shunt current to the externally applied charging or discharging current. E.g., an equivalent shunt current of -1 A practically reduces any applied charging current by 1 A and increases the absolute value of any discharging current by 1 A. Under no load conditions, the battery behaves like we would externally apply a discharge current of -1 A.

$$I_{Shunt} = \left| I_{PCS} - \frac{1}{N_C N_S} \sum_{m=1}^{N_S} \sum_{n=1}^{N_C} I_{Cmn} \right| \tag{2-37}$$

Wherein:

I_{Cmn}	Internal cell current of cell n in stack m	(A)
I_{PCS}	Current, applied by the power conversion system	(A)
I_{Shunt}	Equivalent shunt current	(A)
N_C	Number of cells per stack	(-)
N_S	Number of stacks	(-)

The magnitude of the equivalent shunt current increases quadratically with the number of cells for both designs. The Eqs. (2-38) and (2-39) give the parameters of the fitted curves. We can use this approximation to quickly estimate the equivalent shunt current of a stack with a large number of cells from the measured or simulated equivalent shunt current of a short stack, e.g. with five cells. The geometry factor of design 4.6 is 4.48 times bigger than the one of design 2.1. This ratio reflects in the ratio of shunt current magnitudes as well, which is 4.44. This indicates that the channel geometry factor is reciprocally proportional to the shunt current magnitude. The reason for this is that the channel geometry factor is directly proportional to the ionic resistance of the channel, which by far represents the largest resistance in the equivalent electric network. If we assume the voltage source to be constant, the shunt currents are directly proportional to the ionic resistance of the channel and thus to the channel geometry factor.

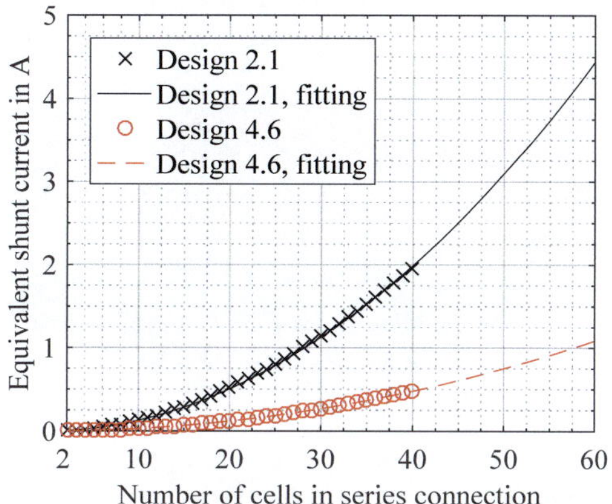

Figure 2-8: Equivalent shunt current during charging over number of cells in a single stack for two designs

Despite the two very different designs, the exponent of the curve fitting equation only varies by 1.7 %. Hence, if the equivalent shunt current of a particular cell design is known for a five-cell stack, the equivalent shunt current of any stack with a larger number of cells can be predicted precisely.

$$I_{\text{Shunt, Design 2.1}}(N_C) = \left(32.4 \cdot 10^{-3} \left(\frac{N_C}{5} \right)^{1.980} \right) A \tag{2-38}$$

$$I_{\text{Shunt, Design 4.6}}(N_C) = \left(7.3 \cdot 10^{-3} \left(\frac{N_C}{5} \right)^{2.014} \right) A \tag{2-39}$$

It is evident that the shunt currents depend on the SoC of the electrolytes, because of their SoC-dependent conductivities. However, the SoC also affects the cell open circuit voltage (OCV) and thus the total cell voltage, which excites the shunt currents. As the battery current affects the cell voltage via the overpotentials as well, it also has an impact on the shunt currents, as shown in Figure 2-9.

In general, the equivalent shunt current is larger for a higher SoC and larger during the charging than during the discharging of the battery. A large charging current boosts the equivalent shunt current while a large discharging current leads to its reduction. This behavior can be traced back to the overpotentials associated with the respective currents. Large charging currents trigger large overpotentials, increasing the cell voltage. Large absolute values of discharging currents trigger large absolute values of overpotentials, decreasing the cell voltage.

Shunt currents also occur for small charging and discharging currents and even if the battery is neither being charged nor discharged. Compared to the externally applied current, which varies from -150 A to 150 A, the equivalent shunt current only varies very little, as shown in Figure 2-9. Hence, shunt currents have a more severe impact on the efficiency if smaller currents are applied to the battery. This is because the ratio of shunt current to load current worsens for smaller currents.

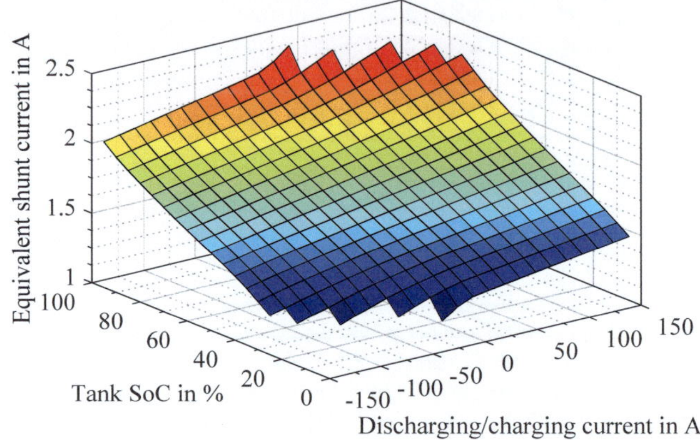

Figure 2-9: Equivalent shunt current versus SoC and current for design 2.1

The electrode area does not affect shunt currents. The comparison of designs 2.1 and 4.6 demonstrates that it is possible to design a larger electrode with a lower equivalent shunt current. The large nominal current of a larger electrode is beneficial as it improves the ratio between externally applied current and equivalent shunt current. Operating a large electrode with a high current density is consequentially an effective measure to eliminate the impact of shunt currents on battery efficiency.

Assuming a current density of 75 mAcm^{-2}, cell design 2.1 carries a current of 150 A. With a shunt current of 1.95 A during the charging of the battery with 150 A at a tank SoC of 50 %, the Coulomb efficiency loss due to shunt currents is 1.3 %-points for a 40-cell stack. For design 4.6, we can double the externally applied current to 300 A. With an equivalent shunt current of 0.48 A, Coulomb efficiency loss is only 0.16 %-points and thus eight times smaller.

2.5.7.2 Internal cell current

In the model, the internal cell current refers to the current which passes the controllable voltage source, as shown in Figure 2-6 on page 27.

Figure 2-10 illustrates the effect of the shunt currents on the individual cell currents in a stack. Herein, the internal cell current distribution on the individual cells is shown for a stack with 10, 20, 30 and 40 cells using the shunt-current-sensitive cell design 2.1.

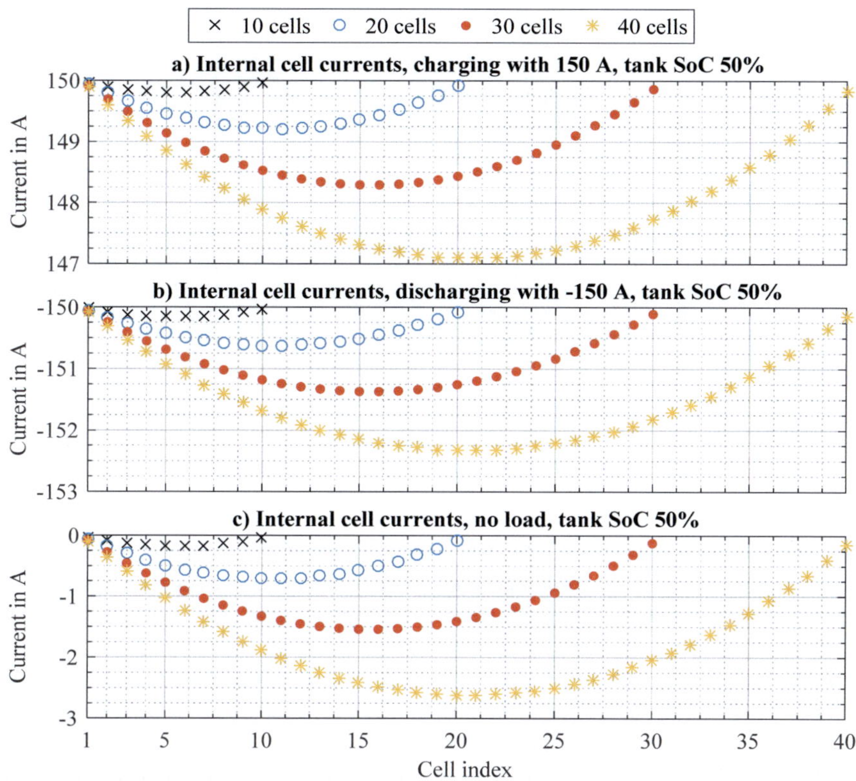

Figure 2-10: Individual cell currents for a stack with 10, 20, 30 and 40 cells of design 2.1

During the charging process, shunt currents lower the internal cell currents, as shown in Figure 2-10 a). Hence, fewer electrons are available in the cells to convert vanadium ions and thus to charge the electrolyte. During the discharging process, shunt currents increase the magnitude of the internal cell current, as shown in Figure 2-10 b). Thus, more 'charged' vanadium ions are required to deliver the desired externally available discharging current. Consequently, the electrolyte is discharged faster than without shunt currents.

As stated before, shunt currents also occur if the battery is neither charged nor discharged, as shown in Figure 2-10 c). If one compares the internal currents of a given cell, e.g. the 20th cell of the 40-cell stack of all three cases (charging, discharging and now load), shown in Figure 2-10 a) to c), we can confirm the dependence of the shunt currents on the externally applied battery current. For a higher charging current, the shunt current magnitude increases, for a larger absolute value of the discharging current, it decreases. This is caused by the increased cell voltage during the charging and the decreased cell voltage during the discharging of the battery.

2.5.8 Multi-stack systems

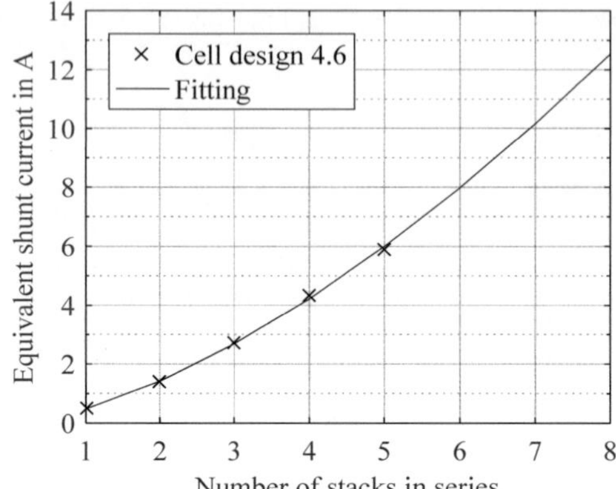

Figure 2-11: Equivalent shunt current over the number of stacks in series connection using cell design 4.6

For manufacturing reasons and the fact that the electrolyte supply via the common manifold is going to evolve inhomogeneously if too many cells are supplied at once, the number of cells per stack is limited. Hence, if a higher voltage than achievable with a single stack is required, we have to connect several stacks electrically in series. Unfortunately, this amplifies the occurrence of shunt currents, as shown in Figure 2-11. Although cell design 4.6 with the largest geometry factor of all studied designs is deployed, for a series connection of five stacks, shunt currents rise to 5.91 A. This corresponds to a loss in Coulomb efficiency of 2.0 %-points for an applied current of 300 A, one way. Again, a curve fitting is deployed to estimate the shunt current losses of the stack series arrangement from the shunt currents of the single stack, as shown in Eq. (2-40). If eight stacks are connected in series, an equivalent shunt current of 12.5 A has to be expected.

$$I_{\text{Shunt, Design 4.6}}(N_{\text{S}}) = \left(0.479 N_{\text{S}}^{1.57}\right)\text{A} \tag{2-40}$$

However, this time, we cannot apply the curve fitting to a general case. This is because shunt currents in multi-stack arrangements strongly depend on the length and the diameters of the external electrolyte piping [9]. For the simulations corresponding to Figure 2-11, the pipe length between the stacks is 1 m. The stacks are connected to the pipe by a tube with a length of 1.5 m. Both tube and pipe have a diameter of 60 mm. If for example the tube diameter is reduced to 30 mm, the equivalent shunt current of five serially connected stacks decreases from 5.91 A to 3.95 A for the considered operation. However, it has to be evaluated if the additional pressure drop due to the decreased tube diameter is tolerable regarding safety and efficiency.

2.6 Open circuit voltage (OCV)

2.6.1 Derivation of the full equations

If we do not attach the battery to any source or load for a longer period of time, it will reach its equilibrium state. The voltage which we then measure between the two electrodes of this battery cell is the so-called open circuit voltage (OCV) [42].

For the VRFB, we can calculate the OCV from the Nernst equation, as shown in Eq. (2-41) [40, 43].

$$E_{OCV} = E^0 + \frac{GT}{F} \ln\left(\frac{a(V^{2+})a(VO_2^+)a^2(H^+)}{a(V^{3+})a(VO^{2+})a(H_2O)}\right) \tag{2-41}$$

Wherein:

a	Activity	(-)
E^0	Cell standard potential	1.255 V
E_{OCV}	Open-circuit-voltage	(V)
F	Faraday constant	96,485 As·mol^{-1}
G	Gas constant	8.314 J(molK)$^{-1}$
T	Temperature	(K)

The activity a describes the deviating behavior of real mixtures from the behavior of ideal mixtures [17]. It is proportional to the product of activity coefficient and ionic concentration, c.

For VRFB modeling, it is most often assumed that the activity coefficients of all ionic species and the water molecules are unity [17]. In this case, the activity of an ionic species is equal to its concentration. As shown in Eq. (2-42), we can calculate the cell OCV, E_{OCVC}, from the ionic concentrations. Herein, c_H denotes the concentration of hydrogen protons.

$$E_{OCVC} = E^0 + \frac{GT}{F} \ln\left(\frac{c_{2C-} \cdot c_{5C+} \cdot c_{HC+}^2}{c_{3C-} \cdot c_{4C+}}\right) \tag{2-42}$$

We can relate the concentration of the vanadium ions to the SoC and the total vanadium concentration, as shown in the Eqs. (2-43) and (2-44).

$$SoC_C = \frac{c_{2C-}}{c_V} = \frac{c_{5C+}}{c_V} \tag{2-43}$$

$$1 - SoC_C = \frac{c_{3C-}}{c_V} = \frac{c_{4C+}}{c_V} \tag{2-44}$$

For hydrogen protons, the correlation shown in Eq. (2-11) on page 17 is used. Finally, we yield Eq. (2-45), in which the OCV only depends on the SoC as a time variant quantity.

$$E_{OCVC} = E^0 + \frac{GT}{F} \ln\left(\frac{SoC_C^2}{(1 - SoC_C)^2}(c_{HC+}(SoC_C = 0) + SoC_C c_V)^2\right) \tag{2-45}$$

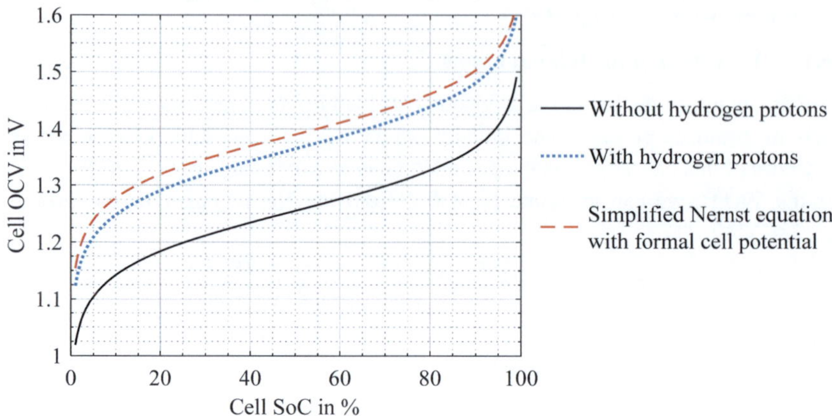

Figure 2-12: Numerical example of different OCV modeling approaches
$(c_{HC+}(SoC_C=0) = 6,000\ \mathrm{molm^{-3}})$

Figure 2-12 shows the OCV derived with different modeling approaches. If we neglect the contribution of hydrogen protons, the simulated OCV is significantly lower, which does not reflect the reality [14, 17, 44]. However, considering hydrogen protons almost exclusively adds an offset to the Nernst equation. This is because the concentration of hydrogen protons in the negative and positive electrolyte is high; mainly due to the dissociation of the sulfuric acid [17]. Hence, the comparatively small number of additional hydrogen protons, which are released on both half-sides during the charging process do not introduce significant OCV variations. Thus, it has become a common procedure to consider the proton concentration to be constant in both electrolytes. Consequently, we can remove the proton concentration in the positive electrolyte from the logarithmic term and add it to the standard cell potential. The sum of the standard cell potential and the proton contribution to the OCV is denoted as the formal cell potential, as shown in Eq. (2-46) [41]. The remaining deviation between the complete and the simplified Nernst equation in Figure 2-12 can be explained by the standard potential in the complete Nernst equation. The standard potential refers to a total vanadium concentration of $1,000\ \mathrm{molm^{-3}}$. In the electrolyte considered in this work, a vanadium concentration of $1,600\ \mathrm{molm^{-3}}$ is used, which increases the OCV. The OCV is also strongly affected by the concentration of the sulfuric acid. A higher sulfuric acid concentration relates to a higher OCV [17].

$$E_{\mathrm{OCVC}} = \widetilde{E}^0 + \frac{GT}{F} \ln\left(\frac{c_{2C-}\cdot c_{5C+}}{c_{3C-}\cdot c_{4C+}}\right) \qquad (2\text{-}46)$$

Wherein:

$\widetilde{E}^0$ Formal cell potential   (V)

In a well-balanced electrolyte at a SoC of 50 %, all four concentrations in Eq. (2-46) are identical and the OCV equals the formal cell potential. Thus, we can measure the formal cell potential of an arbitrary electrolyte composition by measuring the OCV at a SoC of 50 %.

2.7   Ohmic overpotential

In the flow cell, the electric current has to pass several elements, namely bipolar plates, electrodes and membrane as shown in Figure 1-2 on page 6. All these elements impose an ohmic or ionic resistance to the electric current. Additionally, contact resistances exist between the elements. The voltage drop due to the electric and ionic currents through these resistances is the ohmic overpotential.

In [8], the total cell resistance is analytically computed using the geometry of the individual components as well as the specific conductivities of the deployed materials. However, in this work, only the aggregated value of the cell resistance is considered. This is because the work does not target the optimization of individual materials.

$$E_{\text{Ohm}} = I_{\text{C}} R_{\text{C}} = I_{\text{C}} \frac{\psi}{A_{\text{E}}} \qquad (2\text{-}47)$$

Wherein:

ψ	Area specific resistance	(Ωm^2)
A_{E}	Electrode area	(m^2)
E_{Ohm}	Ohmic overpotential	(V)
R_{C}	Cell ohmic resistance	(Ω)

In general, the cell resistance is given as area specific resistance (ASR) in Ωm^2 or Ωcm^2, which we have to divide by the electrode area to derive the cell ohmic resistance, as shown in Eq. (2-47). The ASR allows for a simple comparison between cells of different sizes. In the literature, values between 1.50 Ωcm^2 and 3.13 Ωcm^2 are reported [16, 17, 30]. The ohmic cell resistance is one of the most important quality criteria for the comparison of different cells. In this work, a constant value of 1.50 Ωcm^2 is assumed for all simulations. This is a low but realistic value for state-of-the-art VRFB stacks.

2.8   Activation overpotential

2.8.1   Full equations

The electrochemical reactions, described in Eq. (1-3), require a certain activation energy. This energy demand introduces the so-called activation overpotential. The activation overpotential is described by the Butler-Volmer equation, as shown in Eq. (2-48) for the negative electrode and in Eq. (2-49) for the positive electrode [14, 18]. Note that actually, the ionic concentrations at the electrode surface have to be used in these equations. However, in a lumped-parameter model, these quantities do not exist. Hence, they are replaced by the macroscopic concentrations in the electrolyte within the cell.

Subscript 'An' denotes the anodic transfer coefficient; subscript 'Ca' denotes the cathodic transfer coefficient. Note, that we have to refer the transfer current density to the electrochemical active electrode surface which is much larger than its geometrical area. This is because of the high porosity of the deployed graphite felt.

$$i_{\mathrm{TF}-} = Fk_{\mathrm{RC}-} \cdot c_{2\mathrm{C}-}{}^{\alpha_{\mathrm{An}-}} \cdot c_{3\mathrm{C}-}{}^{\alpha_{\mathrm{Ca}-}} \cdot \left(e^{\alpha_{-\mathrm{An}}FE_{\mathrm{Act}-}/GT} - e^{-\alpha_{\mathrm{Ca}-}FE_{\mathrm{Act}-}/GT} \right) \qquad (2\text{-}48)$$

$$i_{\mathrm{TF}+} = Fk_{\mathrm{RC}+} \cdot c_{4\mathrm{C}+}{}^{\alpha_{\mathrm{An}+}} \cdot c_{5\mathrm{C}+}{}^{\alpha_{\mathrm{Ca}+}} \cdot \left(e^{\alpha_{+\mathrm{An}}FE_{\mathrm{Act}+}/GT} - e^{-\alpha_{\mathrm{Ca}+}FE_{\mathrm{Act}+}/GT} \right) \qquad (2\text{-}49)$$

Wherein:

α	Charge transfer coefficient	(-)
E_{Act}	Activation overpotential	(V)
i_{TF}	Transfer current density	(Am^{-2})
k_{RC}	Electrochemical rate constant	(ms^{-1})

We can convert the transfer current density into the more practical electric cell current I_{C} as shown in Eq. (2-50). The product of the graphite felt specific surface area, s_{F}, with the electrode volume results in the total area which is supposed to be available for the electrochemical reactions within the half-cell.

In the model, the cell current only has one path through the cell. Hence, the values of $i_{\mathrm{TF}-}$ and $i_{\mathrm{TF}+}$ have to be identical.

$$I_{\mathrm{C}} = s_{\mathrm{F}} V_{\mathrm{E}}\, i_{\mathrm{TF}-} = s_{\mathrm{F}} V_{\mathrm{E}}\, i_{\mathrm{TF}+} \qquad (2\text{-}50)$$

Wherein:

s_{F}	Graphite felt specific surface area	(m^{2}m^{-3})
V_{E}	Electrode volume	(m^3)

Applying the Butler-Volmer equation to the VRFB is tainted with uncertainties. This is because the specific surface area of the felt, the charge transfer coefficients and the electrochemical rate constant are not precisely known. Table 2-1 shows an excerpt of published values for the aforementioned parameters. For the specific surface area of the porous graphite electrode, the published values differ by two orders of magnitude. This is in particular noteworthy, as this difference is also present between values, derived experimentally [18, 45, 46].

Table 2-1. Parameters for the computation of the activation overpotential, overview adapted from [46], [1]calculated, [2]measured, [3]estimated

Parameter	Reported values and reference
s_{F} in m^{-1}	$1.62\cdot10^4$ [47][1], $2.3\cdot10^4$ [45][2], $3.5\cdot10^4$ [18][2], $1.2\cdot10^6$ [46][2], $2\cdot10^6$ [22, 48][3]
$\alpha_{\mathrm{An}-}$	0.31 [49], 0.45 [18], 0.5 [48, 50–53]
$\alpha_{\mathrm{Ca}-}$	0.26 [49], 0.45 [18], 0.5 [48, 50–53]
$\alpha_{\mathrm{An}+}$	0.13 [49], 0.5 [48, 50–53], 0.55 [18]
$\alpha_{\mathrm{Ca}+}$	0.14 [49], 0.5 [48, 50–53], 0.55 [18]
$k_{\mathrm{RC}-}$ in ms^{-1}	$4.5\cdot10^{-6}$ [22], $1.7\cdot10^{-7}$ [48, 54], $7.0\cdot10^{-8}$ [18]
$k_{\mathrm{RC}+}$ in ms^{-1}	$6.8\cdot10^{-7}$ [55], $2.5\cdot10^{-8}$ [18], $3\cdot10^{-9}$ [56], $7.6\cdot10^{-9}$ [22]

2.8.2 Simplified equations

For the charge transfer coefficient α, it is widely accepted to assume an equal value of 0.5 for all four different coefficients. This assumption allows for simplifying the Butler-Volmer equation by expressing the double-exponential term as hyperbolic sine, as shown in the Eqs. (2-51) and (2-52).

$$
\begin{aligned}
i_{\mathrm{TF}-} &= Fk_{\mathrm{RC}-}\cdot\sqrt{c_{2\mathrm{C}-}\cdot c_{3\mathrm{C}-}}\cdot\left(e^{^{1}/_{2}FE_{\mathrm{Act}-}/_{GT}} - e^{-^{1}/_{2}FE_{\mathrm{Act}-}/_{GT}}\right) \\
&= 2Fk_{\mathrm{RC}-}\cdot\sqrt{c_{2\mathrm{C}-}\cdot c_{3\mathrm{C}-}}\cdot\sinh\left(\frac{FE_{\mathrm{Act}-}}{2GT}\right)
\end{aligned}
\tag{2-51}
$$

$$
\begin{aligned}
i_{\mathrm{TF}+} &= Fk_{\mathrm{RC}+}\cdot\sqrt{c_{4\mathrm{C}+}\cdot c_{5\mathrm{C}+}}\cdot\left(e^{^{1}/_{2}FE_{\mathrm{Act}+}/_{GT}} - e^{-^{1}/_{2}FE_{\mathrm{Act}+}/_{GT}}\right) \\
&= 2Fk_{\mathrm{RC}+}\cdot\sqrt{c_{4\mathrm{C}+}\cdot c_{5\mathrm{C}+}}\cdot\sinh\left(\frac{FE_{\mathrm{Act}+}}{2GT}\right)
\end{aligned}
\tag{2-52}
$$

Using Eq. (2-50), we can finally convert the Butler-Volmer equation into a form which only uses the macroscopic quantities, available in a lumped-parameter model, as shown in the Eqs. (2-53) and (2-54). The assumption of transfer coefficients equal to 0.5 allows for a direct computation of the activation overpotential out of the cell current, the concentrations of the vanadium ions, the electrochemical rate constant and the specific surface area of the graphite felt.

$$
E_{\mathrm{Act}-} = \frac{2GT}{F}\,\mathrm{arsinh}\left(\frac{I_{\mathrm{C}}}{2s_{\mathrm{F}}V_{\mathrm{E}}Fk_{\mathrm{RC}-}\cdot\sqrt{c_{2\mathrm{C}-}\cdot c_{3\mathrm{C}-}}}\right)
\tag{2-53}
$$

$$
E_{\mathrm{Act}+} = \frac{2GT}{F}\,\mathrm{arsinh}\left(\frac{I_{\mathrm{C}}}{2s_{\mathrm{F}}V_{\mathrm{E}}Fk_{\mathrm{RC}+}\cdot\sqrt{c_{4\mathrm{C}+}\cdot c_{5\mathrm{C}+}}}\right)
\tag{2-54}
$$

2.8.3 Numeric example and critical assessment

Because of the variety of different values for each input parameter, the derived activation overpotential varies significantly, as shown in Figure 2-13. With an assumed ASR of 1.5 $\Omega\mathrm{cm}^2$, the ohmic overpotential is 37.5 mV for a current density of 25 mAcm^{-2} and 112.5 mV for a current density of 75 mAcm^{-2}. With the parameters taken from [18], the activation overpotential exceeds the ohmic overpotential for low and high SoC values. With the parameters presented in [57], the activation overpotential is significantly smaller. Finally, for the parameters given in [22], the activation overpotential practically becomes negligible. The determination of the correct parameters, e.g. the electrochemical rate constant, is still subject to recent research work [58, 59].

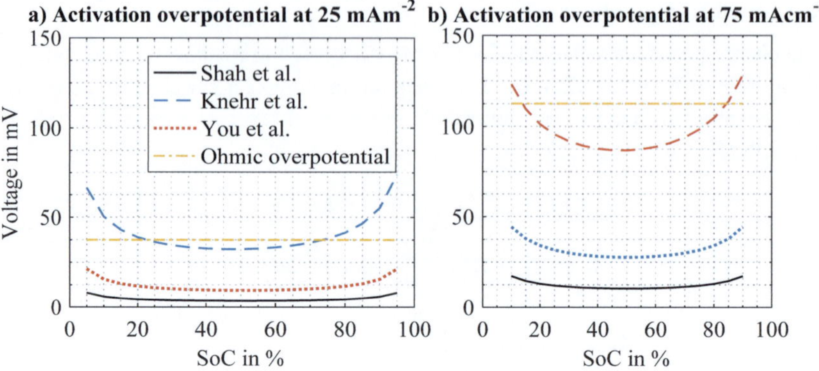

Figure 2-13: Numerical examples of activation overpotential, derived with three different parameter datasets for an electrode height of 4 mm
Specific surface area, transfer coefficient and electrochemical rate constant from Shah et al. [22] (s_F=2·10^6 m^{-1}, α=0.5, k_{RC-}= 1.75·10^{-7} ms^{-1}, k_{RC+}= 3·10^{-9} ms^{-1}), Knehr et al. [18] (s_F=2·10^6 m^{-1}, α_-=0.45, α_+=0.55, k_{RC-}= 7·10^{-8} ms^{-1}, k_{RC+}= 2.5·10^{-8} ms^{-1}) and You et al. [57] (s_F=1.62·10^4 m^{-1}, α=0.5, k_{RC-}= 1.7·10^{-7} ms^{-1}, k_{RC+}= 6.8·10^{-7} ms^{-1})

In this work, the activation overpotential is not considered any further for the following reasons:

1. The variety of values for each parameter makes it impossible to identify the correct parameter set for the considered materials without conducting material research and additional experiments.

2. No spatial variation of any quantity is considered within the cell. For the application of the Butler-Volmer equation, this means that the current transfer density is equal all over the cell. As shown in [51] for the cross-flow direction, this is not the case. The current density close to the current collectors is larger than close to the membrane. In spatially resolved models, this can be considered, and the Butler-Volmer equation applies to a large number of small volume elements to compute the individual overpotential in each one of them. In the lumped-parameter model, the equation describes the total cell volume at once. The error which is introduced by this simplification cannot be determined without extensively comparing the lumped-parameter model to a spatially resolved model.

3. For the purpose of the presented model, the activation overpotential only plays a minor role. As presented above, it is mainly influenced by material parameters, which are not subject to change in this work. Hence, the relative differences between any design variation are not influenced by the activation overpotential. Only the absolute values of the results such as the efficiency might show some deviations. Note that this is only valid, because the temperature is constant for all considerations. Different temperatures would cause fluctuations in the activation overpotential and thus affect the comparability of the results. However, this is not the case in the present work.

4. Finally and most importantly, for state-of-the-art flow battery stacks, the activation overpotential can obviously be neglected, because the model without the activation overpotential shows a good agreement with the experimental data as shown in Section 4.3.3 on page 69. The large surface area and the possible thermal or plasma activation of the graphite felt as well as additional improvements as described in [47, 60] effectively minimize the activation overpotential.

2.9 Overpotential due to deviations between tank OCV and cell EMF

Because of the charging and discharging process, the electrolyte in the cell naturally has a different SoC than the electrolyte in the reservoir. This we can only prevent, if we replace the charged or discharged electrolyte with electrolyte from the tank with an infinite velocity.

During the charging process, the average cell SoC is higher than the tank SoC. During the discharging process, the average SoC of the cell is lower than the tank SoC. The SoC variations reflect in corresponding variations of the voltage, calculated with the Nernst equation. Normally, this voltage is called open circuit voltage (OCV). However, as the cell does not operate under open circuit conditions, technically, this term is not applicable here. Hence, it is denoted as electromotive force (EMF).

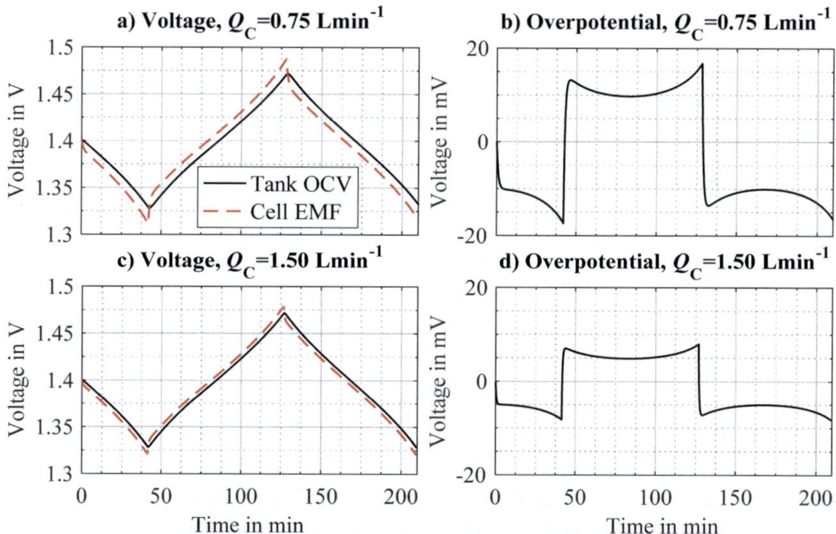

Figure 2-14: Difference between cell and tank OCV for different flow rates for a cycle between tank SoC 20 % and 80 % with 200A

We have to consider the difference between cell EMF and tank OCV as an additional overpotential. This overpotential mainly depends on the ratio between the applied current and the applied flow rate, as shown in Figure 2-14. For a high flow rate of 1.5 Lmin^{-1}, the voltage difference between cell and tank OCV is significantly smaller.

Unfortunately, in general, a high flow rate causes a high pump power demand, which at least partly compensates the efficiency gain due to the reduced overpotential.

2.10 Concentration overpotential

2.10.1 Phenomenon and modeling according to literature

The actual charging and discharging reactions inside a VRFB occur on the surface of the electrode, which normally is a porous graphite felt. The fibers of the graphite felt are very thin and thus offer a large surface area to the electrochemical reactions.

However, the reactants have to reach the surface before the reactions can take place and they also have to get back into the bulk electrolyte after the reaction has taken place. The process of reaching the electrode and getting back into the bulk electrolyte is diffusion-limited [17, 30]. Hence, a diffusion layer evolves in the vicinity of the graphite fibers. In this diffusion layer, the concentration of vanadium ions varies significantly from the concentration within the bulk electrolyte, passing through the cell.

The resulting overpotential is denoted as concentration overpotential, E_{COP}. The Nernst equation can be employed to describe the phenomenon, as shown in Eq. (2-55) [30]. Herein, subscript 'DL' denotes the concentration in the diffusion layer and 'C' denotes the cell concentration of a particular vanadium species.

$$E_{COP} = \frac{GT}{F} \ln\left(\frac{c_C}{c_{DL}}\right) \tag{2-55}$$

To derive the concentration within the diffusion layer, Fick's law can be used to determine the diffusion rate and thus the current density i_{DL} in the diffusion layer, as shown in Eq. (2-56).

$$i_{DL} = F D_{El} \frac{(c_C - c_{DL})}{\delta_{DL}} \tag{2-56}$$

Herein, D_{EL} is the diffusion coefficient of the respective vanadium ion in the electrolyte and δ_{DL} is the thickness of the diffusion layer. Diffusion coefficient over thickness of diffusion layer gives the mass transfer coefficient, k_{MT}, as shown in Eq. (2-57) [30].

$$k_{MT} = \frac{D_{El}}{\delta_{DL}} \tag{2-57}$$

The ions whose concentrations have to be used in Eq. (2-56) are different for the two half-cells and vary with the operation mode. E.g., for the charging process, we have to consider the vanadium ion V^{3+} in the negative half-cell. As V^{3+} ions are reduced to V^{2+} ions during the charging process, the V^{3+} concentration in the diffusion layer is smaller than in the bulk electrolyte. Consequentially, the current density, i_{DL}, is counted positively, which corresponds to our definition, as shown in Section 3.1 on page 57. However, for the discharging process, we have to consider the vanadium ion V^{2+} in the negative half-cell. As V^{2+} ions are oxidized to V^{3+} ions during the discharging process, the V^{2+} concentration in the diffusion layer is smaller than in the bulk or the tank electrolyte. Consequently, the current density, i_{DL}, is again positive. Obviously, the concentration overpotential model does not account for the different signs of the current for the charging and discharging operation. Hence, we have to use the absolute value of the current density in this sub model, as shown in Eq. (2-58). This equation results from combining the Eqs. (2-56) and (2-57) and allows for a calculation of the unknown concentration of vanadium ions within the diffusion layer.

$$c_{\mathrm{DL}} = c_{\mathrm{C}} - \frac{|i_{\mathrm{DL}}|}{F k_{\mathrm{MT}}} \tag{2-58}$$

In Eq. (2-58), we find a mathematical limitation of the presented concentration overpotential model. Obviously, negative concentrations do not exist. Furthermore practically, a concentration is never zero. Hence, we derive condition (2-59) which leads to condition (2-60) for the deployment of the presented model.

$$c_{\mathrm{C}} - \frac{|i_{\mathrm{DL}}|}{F k_{\mathrm{MT}}} > 0 \tag{2-59}$$

$$\frac{|i_{\mathrm{DL}}|}{F k_{\mathrm{MT}} c_{\mathrm{C}}} < 1 \tag{2-60}$$

Finally, the concentration overpotential for both half-cells and both the charging and discharging process can be computed as shown in the Eqs. (2-61) to (2-64). Note that the vanadium ions V^{2+} and V^{3+} have identical diffusion coefficients in the electrolyte, as shown in Section B.2 on page 152. The same is valid for VO^{2+} and VO_2^{+} ions. However, the two diffusion coefficients deviate from each other. Consequently, different mass transfer coefficients have to be considered on the positive and negative half-side.

- Negative half-cell, charging:

$$E_{\mathrm{COPcharging-}} = -\frac{GT}{F} \ln\left(1 - \frac{|i_{\mathrm{DL}}|}{F k_{\mathrm{MT-}} \cdot c_{3C-}}\right), \text{ for } \frac{|i_{\mathrm{DL}}|}{F k_{\mathrm{MT-}} \cdot c_{3C-}} < 1 \tag{2-61}$$

- Negative half-cell, discharging:

$$E_{\mathrm{COPdischarging-}} = -\frac{GT}{F} \ln\left(1 - \frac{|i_{\mathrm{DL}}|}{F k_{\mathrm{MT-}} \cdot c_{2C-}}\right), \text{ for } \frac{|i_{\mathrm{DL}}|}{F k_{\mathrm{MT-}} \cdot c_{2C-}} < 1 \tag{2-62}$$

- Positive half-cell, charging:

$$E_{\mathrm{COPcharging+}} = -\frac{GT}{F} \ln\left(1 - \frac{|i_{\mathrm{DL}}|}{F k_{\mathrm{MT+}} \cdot c_{4C+}}\right), \text{ for } \frac{|i_{\mathrm{DL}}|}{F k_{\mathrm{MT+}} \cdot c_{4C+}} < 1 \tag{2-63}$$

- Positive half-cell, discharging:

$$E_{\mathrm{COPdischarging+}} = -\frac{GT}{F} \ln\left(1 - \frac{|i_{\mathrm{DL}}|}{F k_{\mathrm{MT+}} \cdot c_{5C+}}\right), \text{ for } \frac{|i_{\mathrm{DL}}|}{F k_{\mathrm{MT+}} \cdot c_{5C+}} < 1 \tag{2-64}$$

2.10.2 Mass transfer coefficient

The mass transfer to a carbon or graphite felt is of interest for many applications. Several attempts have been made to identify relevant dependencies [61–63].

Section 2.10 – Concentration overpotential

To transform experimental results, obtained with other redox systems, we can use the dimensionless Sherwood number, which relates to the mass transfer coefficient as shown in Eq. (2-65) [62].

$$Sh = k_{MT}\frac{d_F}{D_{eff}} \tag{2-65}$$

Wherein:

d_F Fiber diameter of the graphite felt   (m)
D_{eff} Effective diffusion coefficient in the electrolyte (m^2s^{-1})

We can calculate the effective diffusion coefficient from the measured diffusion coefficient using the Bruggeman correction, as shown in Eq. (2-66) [18, 64].

$$D_{eff} = D_{El}\varepsilon^{3/2} \tag{2-66}$$

Wherein:

D_{El} Measured diffusion coefficient in the electrolyte (m^2s^{-1})
ε Porosity of the graphite felt   (-)

We can experimentally relate the Sherwood number to the Reynolds number, Re, as shown in the Eqs. (2-67) and (2-68) [61, 62]. In these two references, the mass transfer coefficient towards carbon fiber electrodes is studied using the reduction of $[Fe(CN)_6]^{3-}$ to $Fe(CN)_6]^{4-}$. It is widely assumed that the mass transfer coefficient which is obtained using that reaction is also applicable to the particular vanadium reactions [8].

$$Sh = 7.00Re^{0.40} \tag{2-67}$$

$$Sh = 6.13Re^{0.36} \tag{2-68}$$

We can express the Reynolds number in quantities, which are commonly used to describe a VRFB, as shown in Eq. (2-69).

$$Re = \frac{\rho_{El}v_{El}d_F}{\mu_{El}} \tag{2-69}$$

Wherein

ρ_{El} Electrolyte density (kgm^{-3})
μ_{El} Electrolyte dynamic viscosity   (Pas)
v_{El} Electrolyte velocity (ms^{-1})

If we combine the Eqs. (2-65), (2-67) and (2-69), we derive an actual relation for the mass transfer coefficient that only contains known quantities.

$$k_{MT} = \frac{D_{eff}}{d_F}7.00\left(\frac{\rho_{El}d_F}{\mu_{El}}\right)^{0.40}v^{0.40} \tag{2-70}$$

We can further replace the fluid velocity by the flow rate over the electrode cross sectional area.

$$k_{MT} = \frac{D_{eff}}{d_F}7.00\left(\frac{\rho_{El}d_F}{\mu_{El}}\right)^{0.40}\left(\frac{Q_C}{CSA_E}\right)^{0.40} \tag{2-71}$$

As mentioned before, the diffusion coefficient in the electrolyte is reported to be different for the positive and negative half-cell [55]. Hence, the mass transfer coefficient is also different for the two half-sides. With the two different relations between Sherwood and Reynolds number (Eqs. (2-67) and (2-68)), two possible combinations to describe the mass transfer are yielded. The relevant input parameters can be found in Section B.2 on page 152.

- Negative half-cell, Sherwood to Reynolds number relation (2-67)

$$k_{\mathrm{MT-}} = 1.608 \cdot 10^{-4} \left(\frac{Q_C}{CSA_E} \right)^{0.40} \tag{2-72}$$

- Positive half-cell, Sherwood to Reynolds number relation (2-67)

$$k_{\mathrm{MT+}} = 2.613 \cdot 10^{-4} \left(\frac{Q_C}{CSA_E} \right)^{0.40} \tag{2-73}$$

- Negative half-cell, Sherwood to Reynolds number relation (2-68)

$$k_{\mathrm{MT-}} = 1.322 \cdot 10^{-4} \left(\frac{Q_C}{CSA_E} \right)^{0.36} \tag{2-74}$$

- Positive half-cell, Sherwood to Reynolds number relation (2-68)

$$k_{\mathrm{MT+}} = 2.149 \cdot 10^{-4} \left(\frac{Q_C}{CSA_E} \right)^{0.36} \tag{2-75}$$

The proportional factor of the mass transfer coefficients according to the Eqs. (2-74) and (2-75) is 12 % smaller than according to the Eqs. (2-72) and (2-73). The exponent is 10 % smaller as well. However, as the fluid velocity in practice is always smaller than $1 \ \mathrm{ms^{-1}}$, a smaller exponent corresponds to a larger mass transfer coefficient. This compensates for the smaller proportional factor. In fact, for a realistic fluid velocity in the range of $0.01 \ \mathrm{ms^{-1}}$, the respective coefficients for the negative and the positive half-cell hardly deviate for the two different relations between the Reynolds and the Sherwood number.

2.10.3 Critical assessment of the application in a lumped-parameter model

Despite the fact that the above presented model of the concentration overpotential is validated in [30] for a lumped-parameter model, the application of the method is not unproblematic. The mass transfer coefficient as shown in Eq. (2-72) is most often used to describe the concentration overpotential in both half-cells [16, 19, 21, 30, 51, 57, 65–70]. To the authors best knowledge, no model accounts for the different mass transfer coefficients in the two half-cells.

Another issue concerns the current density in the diffusion layer, i_{DL}, in particular to which area it relates. In [17] it is stated that the concentration overpotential "is caused by the difference in electroactive species concentration between the bulk solution and the electrode surface". The authors of Ref. [30] state that "Concentration overpotential … is created by the concentration gradient between the bulk electrolyte solution and the electrode surface …". So, in both cases, the term electrode surface is used, which may not be mistaken by the electrode area.

Nevertheless, when the concentration overpotential is modeled as mentioned above in lumped-parameter models, the cell current density, i_C, is typically used as the current density in the concentration overpotential model. However, an arbitrary electrode with an area of 2000 cm^2 and a thickness of 4 mm has an electrode surface of 12.96 m^2, if we assume the smallest specific surface area for a graphite felt electrode from Table 2-1 on page 38, which is $1.62 \cdot 10^4$ m^{-1}. Hence, the geometrical electrode area and the electrode surface and thus the macroscopic current density j_C and the microscopic current density j_{DL} deviate by a factor of 65.

Two sample cycles are simulated with the 2000-cm^2 cell to illustrate the issue. The first cycle is carried out between SoC 20 % and 80 %, the second cycle is carried out between cell voltage limits of 1.1 V and 1.7 V. For both cycles, a macroscopic current density of 75 mAcm^{-2} and a flow rate of 1.5 Lmin^{-1} is applied.

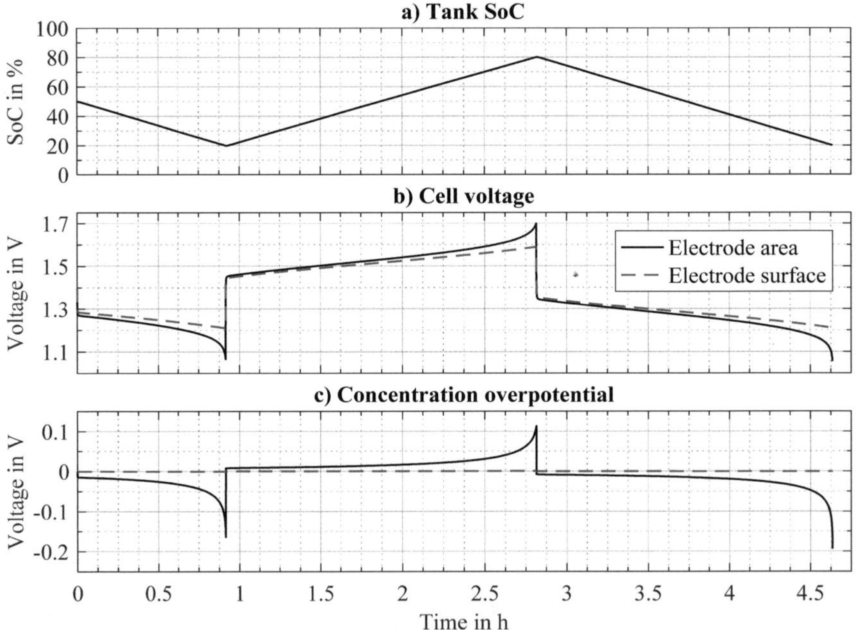

Figure 2-15: Sample cycle between tank SoC 20 % and 80 % for the illustration of the difference between the electrode area and the electrode surface regarding the computation of the concentration overpotential

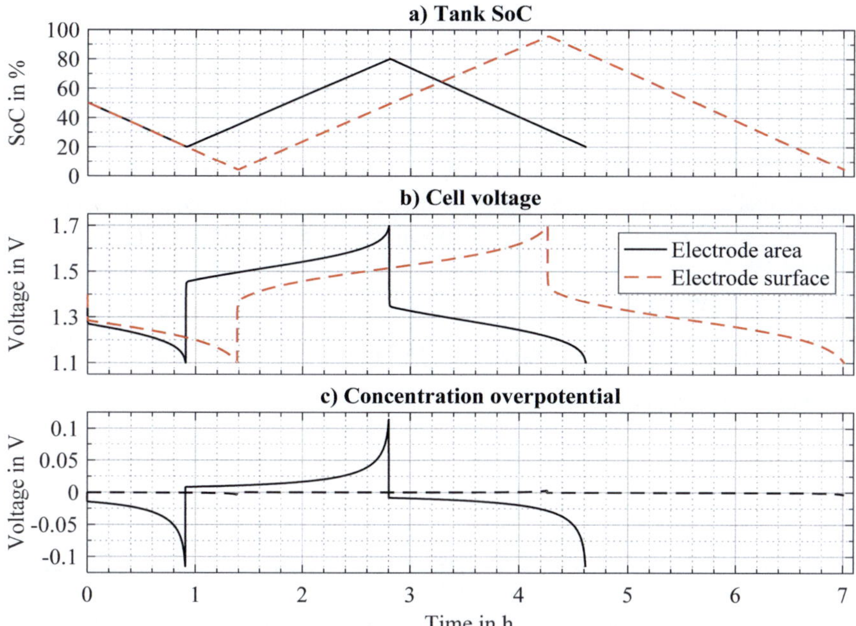

Figure 2-16: Sample cycle between cell voltage limits of 1.1 V and 1.7 V for the illustration of the difference between the electrode area and the electrode surface regarding the computation of the concentration overpotential

If the current density i_{DL} is calculated from the cell current, I_C, using the electrode area, A_E, the concentration overpotential contributes significantly to the rising cell voltage towards the end of the charging process and the dropping voltage towards the end of the discharging process, as shown in Figure 2-15. If the total electrode surface is used to compute the current density in the diffusion layer, the concentration overpotential is negligible.

In the cycle bounded by cell voltage limits, the usage of the electrode area leads to a 35 % shorter cycle time, as shown in Figure 2-16. Again, if the total electrode surface is deployed, the concentration overpotential can be neglected.

There are two main reasons, why neither the electrode area nor the electrode surface is the correct area for calculating the current density.

The first reason is that the exact determination of the electrochemically active surface of a graphite felt is difficult [62]. There might be structures in the electrode, which are blocked at the end. Hence, they contribute to the total electrode surface, but are not available to the electrochemical reactions. In addition, the wettability of the surface as well as the surface structure can affect the effective electrochemical active surface.

The second reason is that the electrochemical activity is not uniformly distributed in a flow cell [51]. Hence, the current density will vary over the electrochemically active surface.

Most likely, the effective area of the diffusion layer is in between the electrode area and the total electrode surface. In this work, experimental results from three flow battery manufacturers, are used to identify the correct area for the calculation of j_{DL}, as shown in Chapter 4 starting on page 65.

2.11 Total cell voltage

The externally measurable cell voltage is the sum of cell OCV, concentration over-potential, and ohmic voltage drop, as shown in the Eqs. (2-76) and (2-77) and in Figure 2-17.

- For the charging process:

$$E_C = E_{OCVC} + E_{COPcharging-} + E_{COPcharging+} + I_C R_C = E_C' + I_C R_C \qquad (2\text{-}76)$$

- For the discharging process:

$$E_C = E_{OCVC} - E_{COPdischarging-} - E_{COPdischarging+} + I_C R_C = E_C' + I_C R_C \qquad (2\text{-}77)$$

The overpotentials have different signs for the charging and discharging process. In this work, the sign definition is identical to the sign definition of the electric current. During the charging process, current and overpotentials are denoted positively. During the discharging process, they are counted negatively.

In the case of the ohmic overpotential, the sign of the electric current defines the sign of the overpotential, because the ohmic resistance is always positive. The value of the concentration overpotential calculated by the Eqs. (2-61) to (2-64) however is always positive. Hence, a negative sign has to be inserted for the discharging process.

a) Charging process

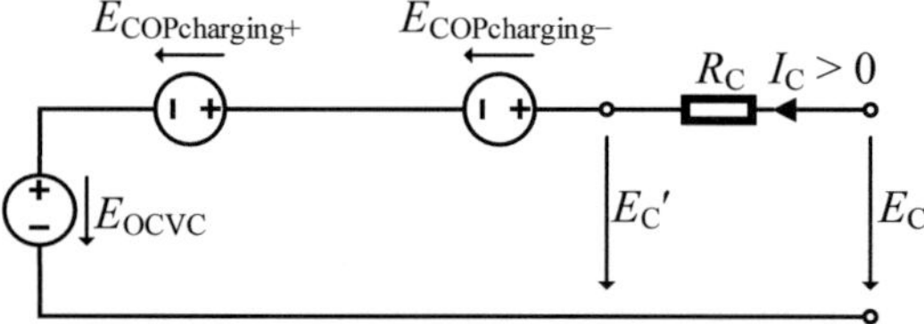

b) Discharging process

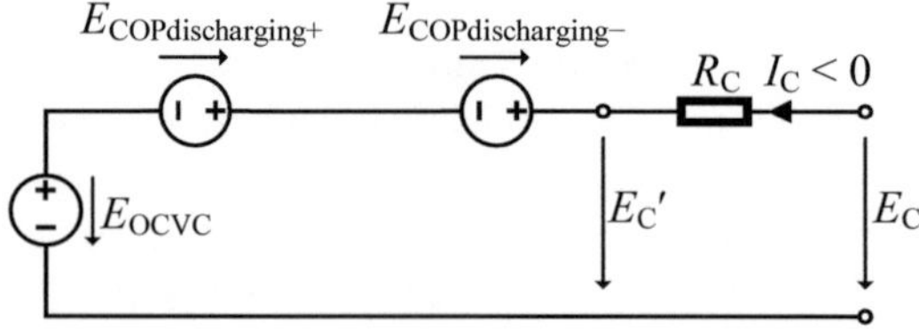

Figure 2-17: Equivalent electric circuit of a single cell used in this work

2.12 Modeling of the hydraulic circuit

The energy demand required for transporting the electrolyte in and out of the stacks, is a significant source of power loss in a flow battery. Furthermore, hydraulic, electric and electrochemical subsystems are closely interrelated. A higher flow rate decreases the concentration overpotential but increases the pump power demand. Long and narrow inlet and outlet channels of the cells reduce shunt currents, but increase the hydraulic resistance and thus again the pump power demand. Hence, the hydraulic circuit compellingly needs to be included into the design and optimization considerations.

The pressure drop in the stack manifolds has to remain small, to enable an equal supply of all connected cells. Consequently, it is neglected in this work. The pressure drop in the cell, which consists of input channel, distribution funnel, porous flow-through electrode, collection funnel and output channel is derived using computational fluid dynamic (CFD) simulations. The cell model set up in ANSYS Maxwell to determine the channel geometry factor is linked to the CFD software ANSYS Fluent. Hence, the geometry does not have to be modelled again.

The elements of the external hydraulic circuit such as pipes, tubes and orifices, namely T-junctions, 90° bends, and sensors, are modelled as follows. Also, a flow rate dependent pump efficiency is considered.

2.12.1 Pressure drop in the porous graphite felt electrode

For the porous flow-through electrode, ANSYS Fluent applies Darcy's law, as shown in Eq. (2-78) [71].

$$\Delta p_E = \frac{\mu_{El} h_E}{\kappa_E w_E \delta_E} Q_C \tag{2-78}$$

Wherein:

Δp_E	Pressure drop in the electrode	(Pa)
μ_{El}	Electrolyte dynamic viscosity	(Pas)
κ_E	Permeability of the porous electrode	(m²)
h_E	Electrode height	(m)
w_E	Electrode width	(m)
δ_E	Electrode thickness	(m)
Q_C	Cell volumetric flow rate	(m³s⁻¹)

The permeability of the porous electrode, κ_E, is derived using the Kozeny-Carman relation, as shown in Eq. (2-79) [71].

$$\kappa_E = \frac{d_F^2}{16 K_{KC}} \frac{\varepsilon_E^3}{(1 - \varepsilon_E)^2} \tag{2-79}$$

Wherein:

d_F	Fiber diameter of the graphite felt	(m)
K_{KC}	Kozeny-Carman constant	4.28
ε_E	Porosity of the graphite felt electrode	0.93

The reciprocal value of κ_E is called viscous resistance and is an input parameter for the cell zone condition of ANSYS Fluent. In [67], K_{KC} is treated as a fitting parameter and determined to be 4.28, corresponding very well to the experiments and computations reported in [71].

2.12.2 Computational fluid dynamic simulations

For the meshing of the geometry, the sweep method is applied to all parts. A face sizing with an edge size of 0.5 mm is assigned to the inlet and the outlet face of the electrode. Thus, the electrode, representing the largest part of the studied problem volume, is discretized more coarsely using rectangles of different sizes. The longer sides of the rectangles are aligned with the fluid flow direction. The rectangles are smaller towards the inlet and outlet zones of the electrode and larger towards the middle of the electrode. All other parts are discretized using cubes of 0.5 mm edge length. Figure 2-18 a) shows the mesh for a 2000-cm^2 cell with a channel width of 20 mm. Herein, the mesh size is increased by a factor of ten for visualization purposes.

After applying the standard initialization, which is computed from the inlet, the problem is solved using up to 500 iterations. The convergence criterion is set to absolute. The absolute convergence criteria are 0.001 for continuity, x-velocity, y-velocity and z-velocity, which represent the standard values of ANSYS Fluent. The option 'Double precision' is activated because the extent of the geometry varies significantly in the three dimensions. Figure 2-18 b) shows the results of the CFD simulation for a 2000-cm^2 cell with a channel width of 20 mm. The fluid velocity in the graphite felt is very homogenous, which is plausible due to its very large hydraulic resistance. In the channel, a laminar flow evolves, with a peak velocity of 0.9 ms^{-1}.

a) Mesh   b) Simulated fluid velocity

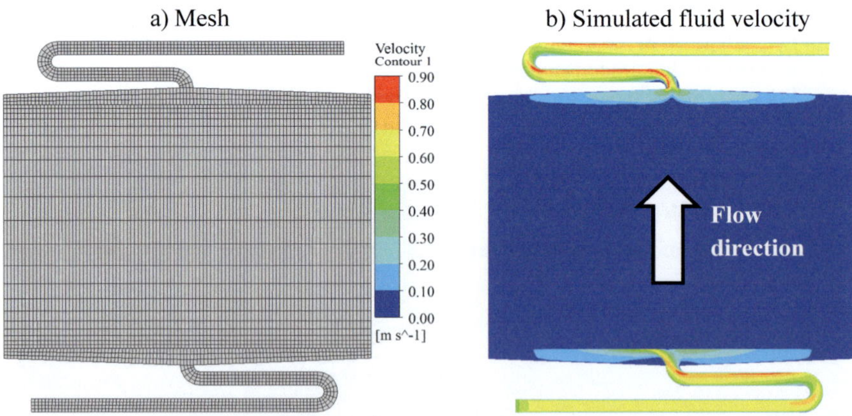

Figure 2-18: CFD model of a 2000-cm^2 cell with a channel width of 20 mm: a) Exemplary meshing with an edge length of 5 mm, b) Fluid velocity in the cell at the nominal flow rate

2.12.3 Implementation into the MATLAB/Simulink model

For an efficient implementation of the CFD simulation results, the following approach is developed. The graphite felt obviously behaves like a linear hydraulic resistance, as shown in Eq. (2-78). The straight channel parts are an additional linear hydraulic resistance if they carry a laminar flow, which is desired to keep the pressure drop low. The orifices in the channel, such as 90° and 180° bends show a quadratic dependence on the flow rate, as shown in Eq. (2-90). Hence, the function shown in Eq. (2-80) is proposed to model the non-linear hydraulic resistance of the cell. Herein, the coefficients β and γ are used, which yet have to be determined.

$$\Delta p_C = \beta Q_C + \gamma Q_C^{\,2} \qquad (2\text{-}80)$$

We can derive the value pair β and γ analytically by using two pairs of values of pressure drop versus flow rate from the CFD simulations. Hence, in principle, only two values have to be determined by means of CFD simulation, which saves computational resources.

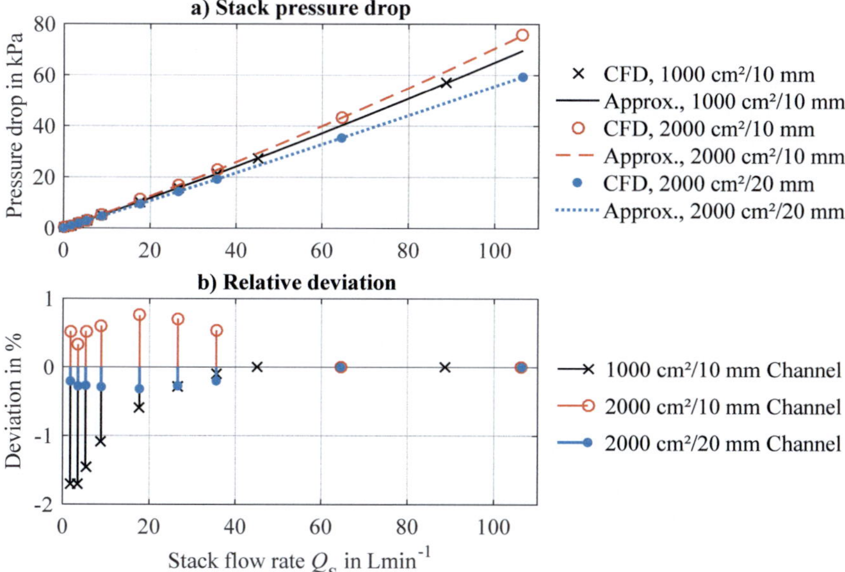

Figure 2-19: Stack pressure drop over flow rate for different electrode areas and different channel widths, results from CFD-simulation and approximation

In the MATLAB/Simulink Simscape library, linear and non-linear hydraulic resistances are available. While the value of β can be used directly, γ has to be converted into an equivalent channel loss coefficient, k_{LCh}, as shown in the Eqs. (2-81) and (2-82). Thereby, the cross-sectional area of the orifice, $A_{Orifice}$, can be set to unity.

$$\Delta p_{\text{Orifice}} = \gamma Q_{\text{C}}^{2} = k_{\text{LCh}} \frac{\rho_{\text{El}}}{2 A_{\text{Orifice}}} Q_{\text{C}}^{2} = \frac{1}{2} k_{\text{LCh}} \rho_{\text{El}} Q_{\text{C}}^{2} \qquad (2\text{-}81)$$

$$k_{\text{LCh}} = \frac{2\gamma}{\rho_{\text{El}}} \qquad (2\text{-}82)$$

The presented approximation represents the results of the time-consuming CFD simulations very precisely, as shown in Figure 2-19. Values for the coefficients β and γ are given in the cell design section, namely in Table 5-4 on page 83.

2.12.4 Pressure drop in pipes

For computing the pressure drop in the pipes, the concept of head loss is used, as shown in Eq. (2-83) [72, 73]. In this concept, the pressure drop Δp is converted into an equivalent gain or loss of height.

$$h_{\text{f}} = (h_{\text{In}} - h_{\text{Out}}) + \left(\frac{p_{\text{In}}}{\rho_{\text{El}} g} - \frac{p_{\text{Out}}}{\rho_{\text{El}} g} \right) = \Delta h + \frac{\Delta p}{\rho_{\text{El}} g} \qquad (2\text{-}83)$$

Wherein:

h_{f}	Head loss	(m)
h	Height	(m)
p	Pressure	(Pa)
ρ_{El}	Electrolyte density	1,354 kgm^{-3}
g	Gravity constant	9.81 N(kg)$^{-1}$

In the common concept of a VRFB, the negative and the positive electrolyte circuits are closed. Hence, Δh is equal to zero over the entire hydraulic circuit. The pressure drop is then only related to the head loss, as shown in Eq. (2-84).

$$\Delta p = h_{\text{f}} \rho_{\text{El}} g \qquad (2\text{-}84)$$

For a pipe with a circular cross-section, the head loss, h_{fP}, is given by Eq. (2-85) [72].

$$h_{\text{fP}} = f \frac{l_{\text{P}}}{d_{\text{P}}} \frac{v_{\text{P}}^{2}}{2g} \qquad (2\text{-}85)$$

Wherein:

f	Friction factor	(-)
l_{P}	Pipe length	(m)
d_{P}	Pipe diameter	(m)
v_{P}	Fluid velocity	(ms^{-1})

Combining the Eqs. (2-84) and (2-85) leads to Eq. (2-86). Therein, the fluid velocity v_{P} in the pipe is expressed as the volumetric flow rate, Q_{P}, over the circular cross-sectional area of the pipe.

$$\Delta p_{\text{P}} = 8f \frac{l_{\text{P}}}{d_{\text{P}}^{5}} \frac{\rho_{\text{El}}}{\pi^{2}} Q_{\text{P}}^{2} \qquad (2\text{-}86)$$

The definition of the friction factor depends on the flow regime [72]. If the Reynolds number is below 2,300, the flow is supposed to be laminar. Thus, the friction factor only depends on the duct shape and the Reynolds number, as shown in Eq. (2-87). For a circular cross-section, the shape factor is 64 [73]. For a Reynolds number beyond 4,000, we assume the flow to be fully turbulent. The friction factor now depends on the Reynolds number and the internal roughness height, ϵ [72]. Between a Reynolds number of 2,300 and 4,000, a linear transition region between the laminar and the turbulent behavior is assumed. The Reynolds number is computed as shown in Eq. (2-88) [73].

$$f = \begin{cases} \dfrac{64}{Re_P}, & \text{for } Re_P \leq 2{,}300 \\[2ex] f_{La} + \dfrac{f_{Tu} - f_{La}}{2{,}300}(Re_P - 2{,}300), & \text{for } 2{,}300 < Re_P \leq 4{,}000 \\[2ex] \left(1.8 \log_{10}\left(\dfrac{6.9}{Re_P} + \left(\dfrac{\epsilon/d_P}{3.7}\right)^{1.11}\right)\right)^{-2}, & \text{for } 4{,}000 \leq Re_P \end{cases} \tag{2-87}$$

$$Re_P = \frac{\rho_{El} v_P d_P}{\mu_{El}} = 4 \frac{\rho_{El} Q_P}{\mu_{El} \pi d_P} \tag{2-88}$$

Wherein:

Re_P Reynolds number in the pipe (-)
f_{La} Friction factor at $Re_P = 2{,}300$ (-)
f_{Tu} Friction factor at $Re_P = 4{,}000$ (-)

2.12.5 Pressure drop due to orifices

The orifices such as bends, T-junctions, tank inlet and outlet, valves, and sensors cause an additional pressure drop. Analogue to the head loss in pipes, the minor loss due to these orifices can be calculated as shown in Eq. (2-89).

$$h_m = k_L \frac{v_{Orifice}^2}{2g} \tag{2-89}$$

Wherein:

h_m Minor loss (m)
k_L Loss coefficient (-)

The loss coefficients of various elements are given in Table 2-2. If the head loss in Eq. (2-84) is replaced by the minor loss according to Eq. (2-89) one yields Eq. (2-90) [72].

$$\Delta p_{Orifice} = 8 k_L \frac{\rho_{El}}{d_{Orifice}^4 \pi^2} Q_{Orifice}^2 \tag{2-90}$$

The tank inlet and outlet are special elements of the hydraulic circuit. In particular in large tanks, an inhomogeneous mixture of the electrolyte in the tank might become a problem. The installation of a stirring device is an option, but causes additional cost and energy loss. Another option is the special design of the tank inlet and outlet to force the mixing within the tank. Most likely, this will cause an additional pressure drop.

However, it is beyond the scope of this work to tackle this problem. Hence, the tank inlet is modeled as a sudden expansion and the tank outlet as a sudden contraction.
In this case, the loss coefficient of the tank inlet is given by Eq. (2-91) [72]. As the exact tank geometry is not known, $CSA_T \gg CSA_P$ is assumed. Therefore, the loss coefficient of the tank inlet is unity.

$$k_{\text{LTIn}} = \left(1 - \frac{CSA_P}{CSA_T}\right)^2 \tag{2-91}$$

Wherein:
CSA_P Pipe cross-sectional area   (m²)
CSA_T Tank cross-sectional area   (m²)

The loss coefficient of the tank outlet is given by Eq. (2-92) [72]. Again, $CSA_T \gg CSA_P$ is assumed. Therefore, the loss coefficient of the tank outlet is 0.42.

$$k_{\text{LTOut}} = 0.42 \left(1 - \frac{CSA_P}{CSA_T}\right)^2 \tag{2-92}$$

Table 2-2: Loss coefficients for various hydraulic elements

Component	k	Ref. / Source
90° bend	0.30	[74]
T-junction, direct flow	0.20	[74]
T-junction, branch flow	1.00	[74]
Tank, inlet	1.00	Eq. (2-91)
Tank, outlet	0.42	Eq. (2-92)

2.12.6 Pump efficiency and pump power

In this work, two pumps apply the desired volumetric flow rate to the two electrolyte circuits. The required electric pump power is calculated with Eq. (2-93) considering a variable, flow rate dependent pump efficiency [73].

$$P_{\text{Pumps}} = 2 \frac{\Delta p_{\text{Total}} Q_{\text{Pump}}}{\eta_{\text{Pump}}(Q_{\text{Pump}})} \tag{2-93}$$

Wherein:
η_{Pump} Pump efficiency   (-)
Δp_{Total} Total pressure difference generated by the pump   (Pa)
P_{Pumps} Electric pump power of both pumps   (W)
Q_{Pump} Pump volumetric flow rate   (m³s⁻¹)

The efficiency of the pump strongly depends on the flow rate, as shown Figure 2-20. The presented curves are derived from datasheets of seven different pumps from three different manufacturers. All efficiencies are derived while pumping water, not VRFB electrolyte. The pump efficiency for pumping electrolyte has not been published yet. The efficiencies of pumps from different manufacturers show a similar trend with a peak efficiency at approximately 60 % of the maximum flow rate. This point is called best-efficiency-point (BEP). However, the efficiencies of different pumps in their BEPs vary strongly. Thus, for this work, an average efficiency is calculated, as shown in Figure 2-21, by taking the arithmetic mean value of the seven normalized efficiency curves, given in Figure 2-20 b).

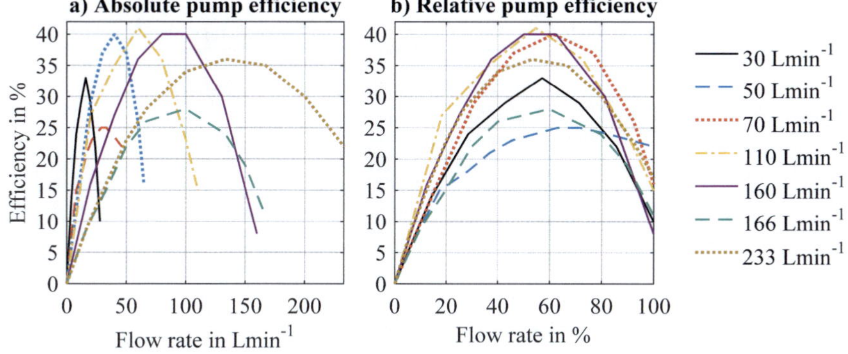

Figure 2-20: Pump efficiency over rated flow rate [75–79]

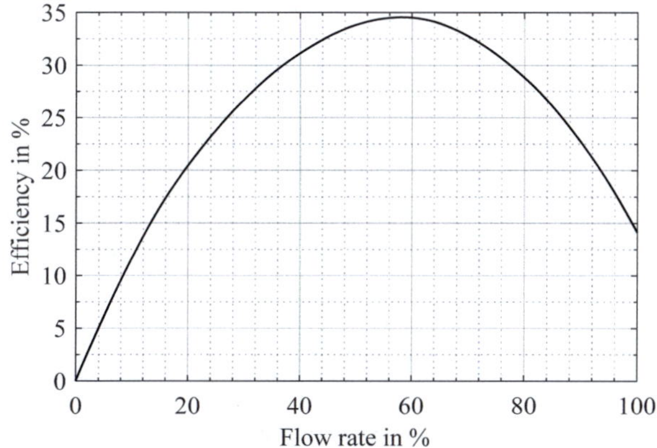

Figure 2-21. Average pump efficiency used in this work over flow rate

2.12.7 Numerical examples

As shown in Figure 2-22, the pressure drop of a pipe with a length of 1 m is in the same order of magnitude as the pressure drop of an orifice with a loss coefficient of one. In both cases, the pressure drop decreases significantly for larger pipe diameters.

Figure 2-23 shows the pressure drop and the electric pump power demand of a single-stack system. For a low flow rate, the pressure drop of the external hydraulic circuit can be neglected compared to the pressure drop of the stack. Even at the highest studied flow rate, the stack still accounts for 89 % of total pressure drop. The pump power demand increases rapidly towards a larger flow rate although the pressure drop approximately increases linearly. This is partly because of multiplying two linearly increasing quantities, namely the flow rate and the pressure drop. However, as shown in Figure 2-21, the efficiency of the pump drops if the delivered flow rate exceeds 60 % of

Section 2.12 – Modeling of the hydraulic circuit

the nominal flow rate. The dropping efficiency additionally amplifies the raise of the pump power towards a larger flow rate.

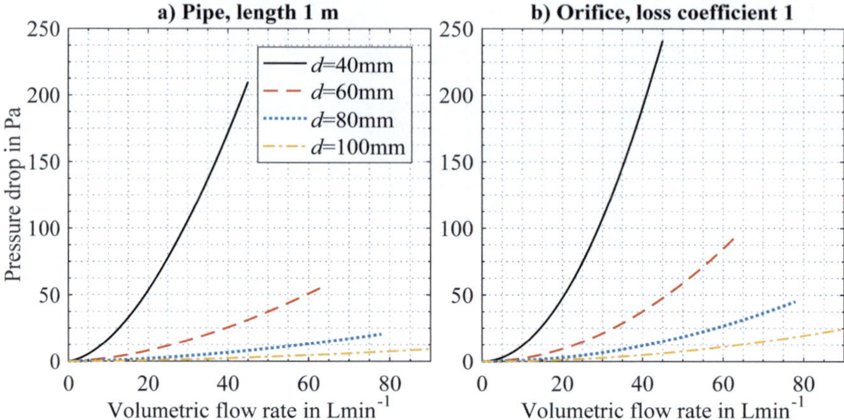

Figure 2-22: Pressure drop over flow rate for a pipe with a length of 1 m and an orifice with a loss coefficient of one and different diameters

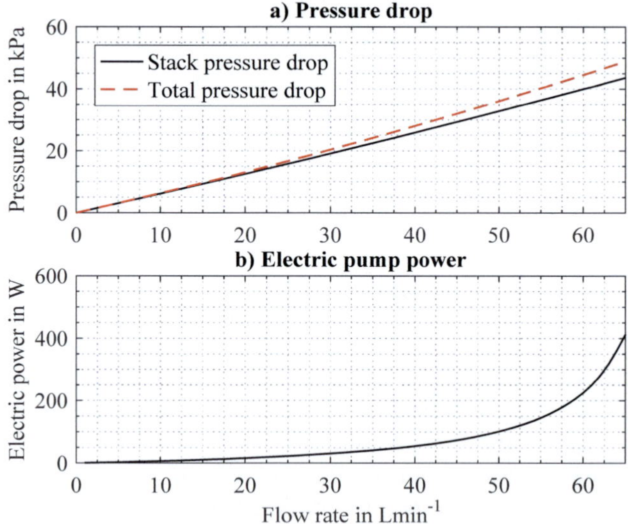

Figure 2-23: Total pressure drop and electric power demand of a sample single-stack system

Chapter 3

Definitions

3.1 Signs of current and power

In this work, the term 'applied current' is used for the current resulting from the voltage difference between the externally applied voltage of the PCS and the internal voltage of the battery. Technically, we apply the voltage and not the current. However, the term 'applied current' shortens and simplifies most of the following explanations.

A charging operation correlates with a positive externally applied current. The power, taken up by the pumps is always counted positively. Hence, the charging process always corresponds to a positive sign of the power. For the discharging of the battery, a negative current is applied externally. Hence, if the pump power does not exceed the absolute value of the power, delivered by the stack(s), the system provides a negative power.

The battery operation is most often evaluated using a full cycle consisting of a consecutive charging and discharging process. In this work, the absolute values of the charging and discharging currents are always equal. Hence, the battery current always refers to the combination of a charging and discharging current with equal absolute values, but different signs. In figures, the denotation charging/discharging current is used, omitting the negative sign of the discharging current. In the text, the term battery current or just current is used to simplify the articulation.

3.2 Nominal operational parameters

The operational parameters, for which the flow battery is designed, are denominated as the nominal operational parameters. The nominal value of a parameter does not correspond to its maximum physical value. If we select a nominal current density of $100\,\mathrm{mAcm^{-2}}$, it is still possible to apply $120\,\mathrm{mAcm^{-2}}$. However, the compliance of all internal parameters within the desired limits is guaranteed only if we do not exceed the nominal operational parameters.

The parameters are closely interrelated to each other. As per definition, we obtain the nominal discharge capacity with the nominal current density and the nominal flow rate. The nominal flow rate is obtained, when the pump fully utilizes its nominal capacity while we measure the nominal pressure drop across the system.

3.3 Capacity

A battery always has two capacities, namely the charging and the discharging capacity. The charging capacity is not a meaningful quantity. If the battery has a low charging efficiency, we need a lot of energy to fully charge it, which corresponds to a large charging capacity. However, this quantity does not differentiate between the energy that is converted into heat during the charging process and the energy that is actually stored in the battery. Consequently, the discharging capacity is exclusively considered in this work. According to the previously made power definition, the discharge capacity is denoted negatively. However, for convenience, the negative sign of the discharge capacity is omitted in the presentation.

3.4 Efficiency definitions

A variety of efficiency definitions exist for battery systems. In this work, it is distinguished between the operation point and the round-trip efficiency. The round-trip efficiencies are deployed in Chapter 6 and Chapter 7 . The operation point efficiency is required for the novel flow rate optimization, presented in Section 8.6 starting on page 127.

3.4.1 Round-trip efficiency

Round-trip efficiencies require a consecutive charging and discharging process. However, different operation modes are possible for both processes. First of all, we can determine either the charging and discharging current or the charging and discharging power. Also, the stack voltage or the tank SoC can be used to indicate if a charging or discharging process is finished.

Table 3-1 gives an overview of the possibilities to determine a round-trip efficiency value. A combination of operation modes is also possible. We can charge the battery with a constant power or a constant current until we reach a certain voltage limit. We can then continue the charging process using a so-called constant voltage phase until we reach a certain SoC limit. During the constant voltage phase, the charging power or current, respectively, is constantly adapted to maintain the cell voltage at the given limit. Naturally, the obtained efficiency values will strongly depend on the chosen operation mode. Combined operation modes are not considered in this work.

Table 3-1 Different efficiency definitions (bold – used in this work)

Operation point efficiency		Round-trip efficiency			
		Limited by SoC		Limited by voltage	
Combination of power and SoC	**Combination of current and SoC**	Constant power	**Constant current**	Constant power	**Constant current**

The term round-trip efficiency is further refined as follows:

- Coulomb efficiency

  The Coulomb efficiency is the ratio of the electric charge, fed to the battery during the charging process, and the electric charge, extracted from the battery during the discharging process, as shown in Eq. (3-1). Herein, I_{PCS} is the externally applied current by the power conditioning system (PCS). The Coulomb efficiency is mainly affected by side reactions such vanadium crossover and shunt currents.

$$\eta_{Col} = -\frac{\int_{t_{Charge}}^{t_{Discharge}} I_{PCS}(t)\,dt}{\int_0^{t_{Charge}} I_{PCS}(t)\,dt} \tag{3-1}$$

- Voltage efficiency

  The ratio between the average stack voltage during the discharging process and the average stack voltage during the charging process, as shown in Eq. (3-2), is called voltage efficiency. The voltage efficiency is mostly affected by the activation, the ohmic and the concentration overpotentials.

$$\eta_{\text{Vol}} = \frac{\int_{t_{\text{Charge}}}^{t_{\text{Discharge}}} E_{\text{S}}(t)\,dt}{\int_0^{t_{\text{Charge}}} E_{\text{S}}(t)\,dt} \frac{t_{\text{Charge}}}{t_{\text{Discharge}} - t_{\text{Charge}}} \tag{3-2}$$

- Energy efficiency

  The ratio between the energy, fed to the battery during the charging process, and the energy, withdrawn from the battery during the discharging process, yields the energy efficiency, as shown in Eq. (3-3). The energy efficiency also corresponds to the product of the Coulomb and the voltage efficiency.

$$\eta_{\text{Ene}} = -\frac{\int_{t_{\text{Charge}}}^{t_{\text{Discharge}}} I_{\text{PCS}}(t) E_{\text{S}}(t)\,dt}{\int_0^{t_{\text{Charge}}} I_{\text{PCS}}(t) E_{\text{S}}(t)\,dt} = \eta_{\text{Col}}\eta_{\text{Vol}} \tag{3-3}$$

- System efficiency

  The Coulomb efficiency, the voltage efficiency and thus the energy efficiency all exclusively refer to the stack. For a functioning flow battery, we have to add auxiliary components such as pumps, PCS, heating, cooling and battery management system (BMS). Losses caused by these components have to be considered as well. However, in this work, no thermal management is taken into account. Further, except for one short consideration in Section 7.4 on page 116, the PCS is excluded from the efficiency computations. Consequently, in this work, the pump power is the only source of auxiliary power loss, as shown in Eq. (3-4). Herein, E_{PCS} is the voltage, which the PCS has to apply in order to enable the current I_{PCS}. The quantity P_{Pumps} is the electric power of both pumps.

  The auxiliary efficiency, η_{Aux}, can be used to compare the effect of the power consumption of pumps and other auxiliary equipment to the voltage, Coulomb and energy efficiency.

$$\eta_{\text{Sys}} = -\frac{\int_{t_{\text{Charge}}}^{t_{\text{Discharge}}} \left[I_{\text{PCS}}(t) E_{\text{PCS}}(t) + P_{\text{Pumps}}(t)\right] dt}{\int_0^{t_{\text{Charge}}} \left[I_{\text{PCS}}(t) E_{\text{PCS}}(t) + P_{\text{Pumps}}(t)\right] dt} = \eta_{\text{Ene}}\eta_{\text{Aux}} \tag{3-4}$$

To take the thermal management into account, a thermal model needs to be included, which exceeds the scope of this work. All other loss sources, such as BMS or PCS affect all cell designs or flow rate control strategies identically. Hence, they only introduce an efficiency offset, but they have no impact on the comparative assessments, conducted in this work.

3.4.2 Challenges in determining the round-trip efficiency

During the determination of the round-trip system efficiency (RTSE), variations in the tank SoC at the beginning and the end of the conducted cycle introduce large errors. This is in particular important when cell voltage limits are used to determine the end of the charging and discharging process. The following example demonstrates the origin of undesired variations in the RTSE and their prevention.

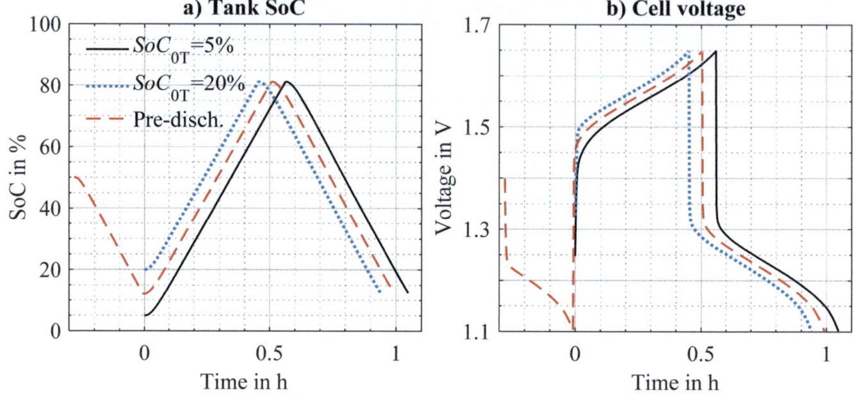

Figure 3-1: Tank SoC and cell voltage over time for a sample cycle with 200 A

Three sample cycles with a stack consisting of 40 2000-cm^2 cells are simulated. A charging/discharging current of 200 A is applied. At the beginning of the first cycle, the tanks have an initial SoC, SoC_{0T}, of 5 %. For the second cycle, it is assumed that the initial tank SoC is 20 %. The third cycle starts with a so-called pre-discharging process from an initial tank SoC of 50 %.

Table 3-2: Efficiencies and discharge capacities of the
three sample cycles shown in Figure 3-1

	SoC_{0T} = 5 %	SoC_{0T} = 20 %	Pre-discharging
Coulomb efficiency	87.4 %	108.4 %	95.3 %
Voltage efficiency	78.6 %	77.7 %	78.2 %
Energy efficiency	68.7 %	84.2 %	74.5 %
System efficiency	67.6 %	82.7 %	73.2 %
Discharge capacity	3.90 kAh	3.90 kAh	3.90 kAh
	4.70 kWh	4.70 kWh	4.70 kWh

Without additional measures, all efficiencies, except for the voltage efficiency, strongly depend on the initial tank SoC, as demonstrated in Table 3-2.

The reason for these variations is the tank energy balance. The tank SoC at the end of the discharging process is independent of the initial value. If the tank SoC at the end of the discharging process is higher than the initial one, a certain part of the energy, fed to the tank during the associated charging process, stays in the reservoir. If the tank SoC at the end of the discharging process is lower than the initial one, energy, which we did not store in the reservoir in the associated charging process, is taken out of the tank. In this case, the Coulomb efficiency might exceed 100 %.

To sum up, if we intend to precisely state on the round-trip efficiency, the initial tank SoC must be equal to the value at the end of the discharging process. In the case of cycles which are limited by SoC limits, this is fulfilled per definition. However, in this case, the SoC of the stack(s) at the beginning and the end of the cycle also affects the RTSE, but to a lower extent [9]. If the cell voltage limits the cycles, it is proposed to conduct the presented pre-discharging process before the actual cycle starts. Note that this process does not relate to the conditioning of the electrolyte. The pre-discharging process starts from a tank SoC of 50 %. It is carried out with the same discharging current or discharging power, as the relevant discharging process and with the same flow rate control strategy. When the lower SoC or voltage limit is reached during the pre-discharging process, the charging process and thus the actual cycle starts, as shown in Figure 3-1. The lower SoC or voltage limit can be chosen to match the purpose of the experiment.

3.4.3 Operation point efficiency

Although we can easily determine the round-trip efficiency in simulated and real systems, it faces some drawbacks. If the efficiency at lower currents and power is of interest, round-trips are very time-consuming. Furthermore, the influence of design or operational parameters, such as the flow rate, is only evaluated in an integral manner over the SoC range used during the cycle. We cannot evaluate the individual contribution of an individual SoC value to the overall round-trip efficiency. Hence, parameters cannot be adapted to specific requirements of an individual SoC value, e.g., in the middle or at the end of the charging and discharging process. Consequently, a concept of operation point efficiency is introduced in [16], which determines the system efficiency in any operation point defined by tank SoC and charging or discharging current.

To determine the operation point efficiency, tank SoC is kept constant. This corresponds to assuming an infinite large tank volume of the positive and negative electrolyte tanks. In general, the instantaneous flow battery efficiency is defined as shown in Eq. (3-5). Herein, P_T is the electrochemical tank power and P_{Sys} is the system power. Both quantities are introduced in the next sections. In the following, the equations are provided for a single-stack system.

$$\eta_{Sys} = \begin{cases} {P_T}/{P_{Sys}}, & \text{for charging} \\ {P_{Sys}}/{P_T}, & \text{for discharging} \end{cases} \tag{3-5}$$

3.4.3.1 Electrochemical tank power P_T

In order to calculate the electrochemical tank power, first, the equivalent electric tank current is computed. By multiplying the ionic flux, J, of V^{2+} and VO_2^+ ions with

Faraday's constant, it can be interpreted as an electric current, as shown in Eq. (3-6) for the negative tank and in Eq. (3-7) for the positive tank. As identical flow rates for each cell are assumed, the flow rate is placed in front of the summation.

$$I_{\text{TIn}-} = FQ_C \sum_{k=1}^{N_C} c_{2Ck-} \tag{3-6}$$

$$I_{\text{TIn}+} = FQ_C \sum_{k=1}^{N_C} c_{5Ck+} \tag{3-7}$$

By multiplying the equivalent tank current with the open circuit voltage (OCV) of the corresponding tank (positive or negative), the equivalent electrochemical tank power, P_T, is computed. It is split into the input tank power, P_{TIn}, shown in Eq. (3-8) and the output tank power, P_{Tout}, shown in Eq. (3-9). We can compute the tank output power more easily as there is only one ionic concentration of interest per positive and negative tank.

$$P_{\text{TIn}} = E_{\text{OCVT}-} \cdot I_{\text{TIn}-} + E_{\text{OCVT}+} \cdot I_{\text{TIn}+} \tag{3-8}$$

$$P_{\text{TOut}} = FQ_T(c_{2T-} \cdot E_{\text{OCVT}-} + c_{5T+} \cdot E_{\text{OCVT}+}) \tag{3-9}$$

To derive the OCV of the negative and the positive electrolyte in the tank, the Nernst equation is applied, as shown in Eq. (3-10). In [41], formal half-cell potentials of 0.207 V and 1.182 V are measured for the negative half-cell and positive half-cell, respectively.

$$\begin{pmatrix} E_{\text{OCVT}-} \\ E_{\text{OCVT}+} \end{pmatrix} = \begin{pmatrix} 0.207\ \text{V} \\ 1.182\ \text{V} \end{pmatrix} + \frac{GT}{F} \begin{pmatrix} \ln\left(c_{2T-}/c_{3T-} \right) \\ \ln\left(c_{5T+}/c_{4T+} \right) \end{pmatrix} \tag{3-10}$$

The actual electrochemical tank power is the difference between the input and output tank power, as shown in Eq. (3-11).

$$P_T = |P_{\text{TIn}} - P_{\text{TOut}}| \tag{3-11}$$

3.4.3.2 System power P_{Sys}

As mentioned before, this work considers the pump power as the only external source of losses. A positive system power corresponds to the charging operation. Thus, power is taken from the grid. The pump power is always positive and always taken from the grid. Consequently, the sign of the pump power does not change for the discharging operation.

$$P_{\text{Sys}} = E_{\text{PCS}} I_{\text{PCS}} + P_{\text{Pumps}} \tag{3-12}$$

3.4.3.3 Sample operation point simulation

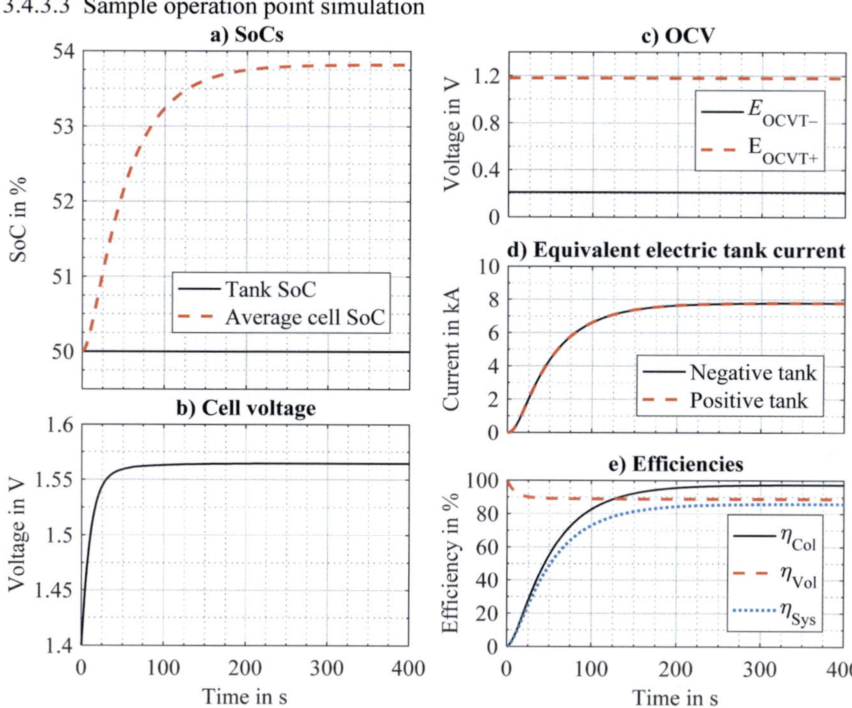

Figure 3-2: Internal quantities for the sample operation point

To illustrate the unusual quantities above, a sample operation point using cell design 2.1 (electrode area 2000 cm², channel length 697 mm and channel width 20 mm) is studied. In the sample operation point, tank SoC is fixed at 50 % and a charging current of 200 A is applied, as shown in Figure 3-2.

Flow rate and charging current are softly started using a first order low-pass filter with a time constant of ten seconds. Concerning the flow rate, this time constant also reflects the time constant of the hydraulic circuit. Consequentially, this filter is active all the time, whereas the current signal is only filtered during the start-up phase.

The cells themselves introduce another time constant that can roughly be estimated by dividing the stack electrolyte volume per half-side by the applied flow rate. In this case, the stack volume per half-side is 32 L. The applied flow rate is 40 Lmin⁻¹. Thus, the estimated stack time constant is 48 seconds. After a simulation time of 400 seconds, the system reaches its steady state. The cells reach an average internal SoC of 53.8 %. Naturally, tank SoC and thus tank OCV does not change during the operation point simulation.

At the beginning of the simulation, no current is fed to the tank, as shown in Figure 3-2 d). First, the cells themselves take up the electric charge which is injected by the applied charging current. Delayed by the stack's time constant that depends on the applied flow rate, the tank charging process starts. In steady state, 7.786 kA are fed to the positive tank, and 7.784 kA are fed to the negative tank. In the ideal case, a charging current of 200 A would be applied to all 40 cells. Hence, the theoretical charging current is 8 kA. Thus, the average Coulomb efficiency in this operation point is 97.3 %.

The voltage efficiency of an operation point is the ratio between tank OCV and cell voltage. Tank total OCV is 1.39 V, while cell voltage is 1.56 V. Thus, the voltage efficiency in this operation point is 89.2 %.

Multiplied with each other, Coulomb and voltage efficiency gives the energy efficiency, which is 86.2 %.

The system efficiency additionally takes into account the power required for pumping the electrolyte. In this case, the system efficiency is 86.1 %. Note that the pump power is only 92.6 W for a flow rate of 40 Lmin^{-1}. Compared to the charging power of 12.5 kW, the pump power is almost negligible. This is typical for operating a flow battery around a SoC of 50 %. The pump power increases strongly for the charging operation at high SoCs and the discharging operation at low SoCs.

Chapter 4

Validation of the stack voltage model

4.1 Available experimental data and confidentiality

It is beyond the scope of this work to build prototypes and to conduct experiments. Thus, VRFB manufacturers are required to supply experimental data for the validation of the battery model. The usage of experimental data obtained with state-of-the-art VRFB stacks is an excellent opportunity to prove the validity of the developed model. However, confidentiality issues are inevitable when cooperating with industrial partners. Naturally, the manufacturers are not interested in having their battery stack compared to their competitors in public. Hence, several measures are taken to guarantee for the confidentiality of the received experimental data:

- Cell and stack parameters are not disclosed.

- The displayed stack voltage is normalized as shown in Eq. (4-1).

$$E_{\text{Normalized}} = \frac{E_{\text{Absolute}} - \min(E_{\text{Experiment}})}{\max(E_{\text{Experiment}}) - \min(E_{\text{Experiment}})} \cdot 100\ \% \qquad (4\text{-}1)$$

Wherein:

$E_{\text{Normalized}}$	Normalized stack voltage	(%)
E_{Absolute}	Absolute measured or simulated stack voltage	(V)
$\min(E_{\text{Experiment}})$	Minimal measured stack voltage during the experiment	(V)
$\max(E_{\text{Experiment}})$	Maximal measured stack voltage during the experiment	(V)

- The three manufacturers are denominated as manufacturer A, B and C. Throughout this section, this denomination is not consistent but changes repeatedly. Although this makes it harder to follow the results and the conclusions, it is inevitable to make it impossible to correlate the shown data to a particular manufacturer.

In total, three manufacturers agreed to deliver experimental data for the model validation. All experiments are conducted in test rigs. The operated stacks comprise between 20 and 40 cells, each with an active cell area of several 100 cm². All stacks have a multi-kW power rating.

In the following, a number of successive cycles under the same operational conditions (e.g. currents, power, flow rate or electrolyte pressure) is called experiment. Each experiment comprises three successive cycles. Each manufacturer provided at least three different experiments.

The experiments of the different manufacturers differ significantly from each other:

- Constant current and constant power cycles are conducted.
- Constant flow rates, equal for the negative and the positive half-side as well as constant pressures across the stack are applied.
- Experiments are conducted with balanced and unbalanced electrolyte.

The heterogeneity of the received data makes it hard to track down possible sources of model inaccuracy. However, if it is possible to match the experimental data with the model, it is valid for a broad range of different operational conditions and different designs.

4.2 Validation methodology

4.2.1 Simulation with literature model

For the stacks of all three manufacturers, the model as presented so far, using Eq. (2-72) on page 45 for both, the negative and the positive half-cell mass transfer coefficient, does not match the experimental results of any of the manufactures, as shown in Figure 4-6 to Figure 4-8 on pages 73 to 74. In all cases, the simulated stack voltage during the charging process is significantly higher than measured during the experiment. During the discharging process, the simulated stack voltage is too low. It is obvious that the deviations are largest at the end of the charging and the discharging process. In the model, the increase of the voltage at the end of the charging process as well as the decrease of the voltage at the end of the discharging process is mainly caused by the concentration polarization. The identified deviations indicate an over-estimation of the concentration overpotential by the model.

4.2.2 Quality criteria of a battery model

4.2.2.1 Root-mean-square error

The most obvious quality criterion is the root-mean-square-error (RMSE) between the measured and the simulated stack voltage, as shown in Eq. (4-2).

$$RMSE = \sqrt{1/N_{\text{Datapoints}} \sum_{n=1}^{N_{\text{Datapoints}}} \left(E_{\text{Experiment}}(n) - E_{\text{Simulation}}(n) \right)^2} \qquad (4\text{-}2)$$

Wherein:

E	Stack voltage	(V)
$N_{\text{Datapoints}}$	Number of data points	(-)
$RMSE$	Voltage root-mean-square-error	(V)

For each experiment k of manufacturer m, an individual RMSE value is computed. The maximum of all k values is taken to state on the overall match between model and stack design of the manufacturer.

4.2.2.2 End-of-charge and end-of-discharge voltage

When cycles are carried out without using constant voltage phases, the stack voltage shows a distinct maximum during the charging process and a distinct minimum during the discharging process, as shown exemplarily in Figure 4-9 and Figure 4-10. As laid out in Section 8.3 on page 121, the match of these two characteristic values is important when simulating the battery using cell voltage boundaries.

A low RMSE value does not necessarily imply a good match of the end-of-charge and the end-of-discharge voltage. This is because the RMSE is dominated by the comparatively long period of time within the cycle, in which the voltage does not change significantly. Therefore, it is possible for a model to obtain a low RMSE, but still to have a significant deviation in the end-of-charge and/or end-of-discharge voltages.

The deviation between the measured and the simulated end-of-charge and end-of-discharge voltage is defined as shown in Eq. (4-3). The maximum deviation over all cycles is taken as the deviation for the particular experiment.

$$\Delta E_{\text{Charge},k,l} = \frac{\left| E_{\text{Experiment}}\left(t_{\text{Charge},k,l}\right) - E_{\text{Simulation}}\left(t_{\text{Charge},k,l}\right) \right|}{E_{\text{Experiment}}\left(t_{\text{Charge},k,l}\right)} \cdot 100\ \%$$

$$\Delta E_{\text{Discharge},k,l} = \frac{\left| E_{\text{Experiment}}\left(t_{\text{Discharge},k,l}\right) - E_{\text{Simulation}}\left(t_{\text{Discharge},k,l}\right) \right|}{E_{\text{Experiment}}\left(t_{\text{Discharge},k,l}\right)} \cdot 100\ \%$$

$$(4\text{-}3)$$

Wherein:

k	Counting index for experiments	(-)
l	Counting index for cycles within an experiment	(-)
$t_{\text{Charge},k,l}$	Time when the lth charging process of the kth experiment is finished	(s)
$t_{\text{Discharge},k,l}$	Time when the lth discharging process of the kth experiment is finished	(s)

4.2.2.3 Combined quality criterion

The criteria $RMSE_k$, $\Delta E_{\text{Charge},k}$ and $\Delta E_{\text{Discharge},k}$ can each serve as an individual quality criterion for the match of model and reality. However, it is preferable to obtain low values for all three criteria. Hence, a combined quality criterion taking into account the values of $RMSE_k$, $\Delta E_{\text{Charge},k}$ and $\Delta E_{\text{Discharge},k}$ is proposed, as shown in Eq. (4-4). Therein, the arithmetic mean of all three quantities is computed, with a double weight on $RMSE_k$.

$$q_k = \frac{1}{4}\left(2\,RMSE_k + \Delta E_{\text{Charge},k} + \Delta E_{\text{Discharge},k}\right) \tag{4-4}$$

In two cases, the manufacturer deployed a constant voltage phase during the charging process. In these cases, only the discharging process shows the aforementioned distinct minimum in the stack voltage. The combined quality criterion is then defined as shown in Eq. (4-5).

$$q_k = \frac{1}{3}\left(2\,RMSE_k + \Delta E_{\text{Discharge},k}\right) \tag{4-5}$$

The combined quality criterion is derived for each experiment. The largest value for all experiments of one particular manufacturer states on the overall match between model and stack of this manufacturer.

4.3 Determination of the electrochemical active electrode area

4.3.1 Model adaption

According to the critical assessment of the concentration overpotential model in Section 2.10.3 on page 45, it is not surprising that the model deviates substantially from reality. To improve the model, individual mass transfer coefficients are now considered for negative and positive half-cell. Further, as the deviations can also be explained by the lack of clarity regarding the area with which the current density has to be computed in the concentration overpotential model, a method for deriving the correct area from the experimental results is presented.

For the negative and positive half-cell, the mass transfer coefficient as shown in the Eqs. (2-72) and (2-73) is deployed. In Eq. (4-6), the electrode area, A_E, is used for calculating the current density i_{DL} from the cell current I_C. The scaling factor K is introduced to adapt the electrode area to the actual electrochemical active electrode surface. A scaling coefficient of larger than one implies a larger electrochemical active electrode surface than the electrode area itself and vice versa. Of course, the former is expected to be observed. For each manufacturer, the scaling factor is determined using a parameter sweep.

$$i_{DL} = \frac{|I_C|}{KA_E} \tag{4-6}$$

Wherein:
K Scaling factor for model adaption   (-)

4.3.2 Model preparation

The hydraulic subsystem and the computation of the vanadium ion concentrations is excluded in this part of the work. Either time dependent concentrations of the vanadium ions are supplied directly by the manufacturer, or they are computed using the supplied time dependent tank SoC and the total vanadium concentration. In the latter case, an ideally balanced electrolyte is assumed, as shown in the Eqs. (4-7) and (4-8).

$$\begin{pmatrix} c_{2T-} \\ c_{3T-} \\ c_{4T-} \\ c_{5T-} \end{pmatrix} = \begin{pmatrix} c_V SoC_T \\ c_V(1 - SoC_T) \\ 0 \\ 0 \end{pmatrix} \tag{4-7}$$

$$\begin{pmatrix} c_{2T+} \\ c_{3T+} \\ c_{4T+} \\ c_{5T+} \end{pmatrix} = \begin{pmatrix} 0 \\ 0 \\ c_V(1 - SoC_T) \\ c_V SoC_T \end{pmatrix} \tag{4-8}$$

As the model does not have to compute tank SoC and vanadium concentrations, vanadium crossover and shunt currents are excluded. This is valid because the impact of these mechanisms on cell voltage is negligible for a well-designed single stack as used by all three manufacturers. Without the presence of shunt currents, all cells of the stack behave exactly identically because an equal supply with electrolyte is assumed for all cells. Hence, it is possible to reduce the stack model to a single-cell model and to derive the stack voltage by scaling the cell voltage with the number of cells per stack. This significantly reduces the demand in computational time. Thus, a large number of parameter variations can be carried out in a reasonable amount of time. This is helpful for deploying finely-resolved parameter sweeps and optimization algorithms to determine missing parameters or to adapt questionable parameters to the reality.

4.3.3 Simulation results

4.3.3.1 Literature model

The derived quality criteria confirm that combining Eq. (2-72) as mass transfer coefficient and using the electrode area, A_E, for calculating the current density, is inappropriate. In terms of RMSE, the average deviations exceed 400 mV on a cell level. Regarding the end of discharge voltage, the deviations range up to 116 %, as shown in Table 4-1 and Figure 4-1 to Figure 4-5.

Table 4-1: Scaling factor and quality criteria for the model using the literature mass transfer coefficient ('Lit.') and the correct mass transfer coefficients for both half-cells ('Val.')

	Manuf. A		Manuf. B		Manuf. C	
	Lit.	Val.	Lit.	Val.	Lit.	Val.
Scaling factor K	1	7.06	1	1.44	1	3.31
q_{max} in mV	752	24	421	28	610	20
$RMSE_{max}$ in mV	491	17	109	10	408	1
$\varnothing RMSE$ in mV	437	15	31	8	373	1
$\Delta E_{Discharge,max}$	116 %	3 %	92 %	7 %	92 %	4 %

4.3.3.2 Individual adaption to every manufacturer

With the proposed scaling factor, the model can be adapted individually to the three studied systems. After the adaption, the model shows an excellent accuracy with RMSE values below 1 mV and acceptable deviations in the end-of-discharge voltage, as shown in Table 4-1 and Figure 4-1 to Figure 4-5. The derived scaling factors range from 1.44 to 7.06, which means that up to 7.06 times the electrode area is available to the electrochemical reactions.

However, unfortunately but not unexpected, it is not possible to use the same scaling factor for all manufacturers. As not all design parameters are provided by the manufacturers, the origin for the different factors cannot be determined in this work. Possible reasons include but are not limited to different materials, different electrode pre-treatments or different compression factors.

4.3.3.3 Average scaling factor

For the following examinations, the arithmetic mean of the scaling factor of manufacturer B and manufacturer C is taken as scaling factor, which is 2.38. The scaling factor of manufacturer A is significantly larger. As the reason for this cannot be identified at this point, it is excluded from further considerations.

As shown in Figure 4-6 to Figure 4-10, the adapted model with a scaling factor of 2.38 correlates significantly better to the measurement data of two manufacturers than the approach commonly used in the literature approach. Even for the third manufacturer, whose scaling factor is excluded from further considerations, the match of the adapted model with a scaling factor of 2.38 is substantially better than the literature model, shown in Figure 4-7 and Figure 4-8. However, in particular towards the end of the discharging process, the simulated voltage is still significantly too low after the adaption of the electrode area with a factor of 2.38. The yielded quality criteria confirm that the deviations between model and reality are significantly smaller with the adapted model using the average scaling factor, as shown in Table 4-2. However, in particular for one manufacturer, significant deviations remain.

Table 4-2: Quality criteria for the model using the average scaling factor of 2.38

	Manuf. A	Manuf. B	Manuf. C
q_{max} in mV	591	37	214
$RMSE_{max}$ in mV	250	12	65
$ØRMSE$ in mV	236	8	62
$\Delta E_{Discharge,max}$	116%	7.8 %	47 %

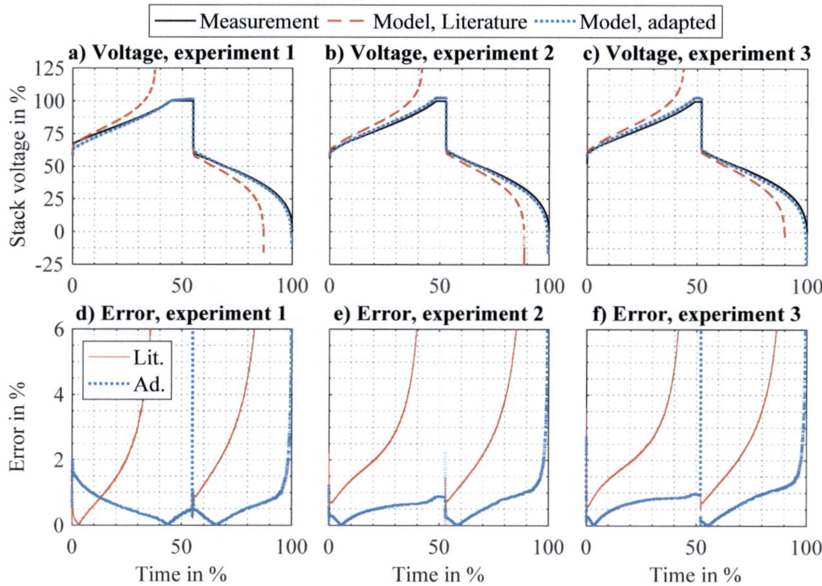

Figure 4-1: Validation with experimental data of manufacturer A with initial and adapted model, **individual scaling factor** for the manufacturer

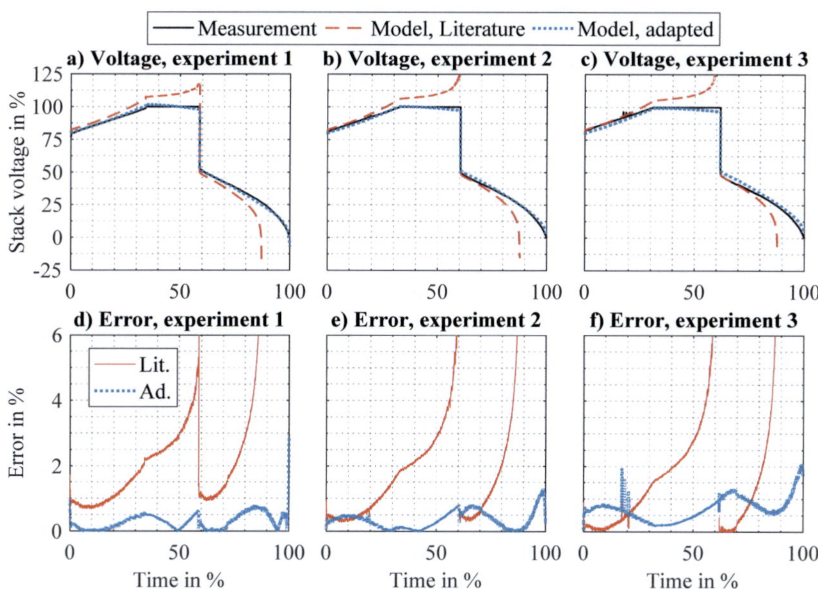

Figure 4-2: Validation with experimental data of manufacturer B with initial and adapted model, **individual scaling factor** for the manufacturer, experiments $1-3$

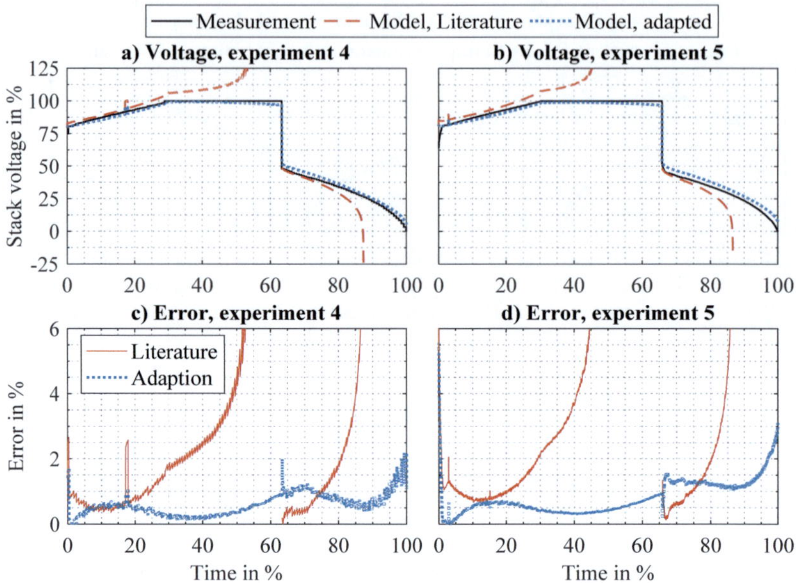

Figure 4-3: Validation with experimental data of manufacturer B with initial and adapted model, **individual scaling factor** for the manufacturer, experiments 4 – 5

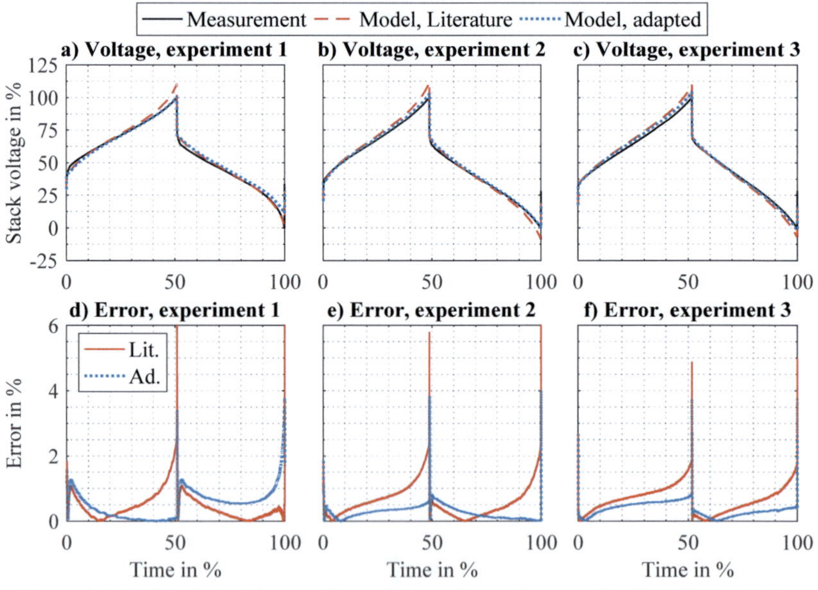

Figure 4-4: Validation with experimental data of manufacturer C with initial and adapted model, **individual scaling factor** for the manufacturer, experiments 1 – 3

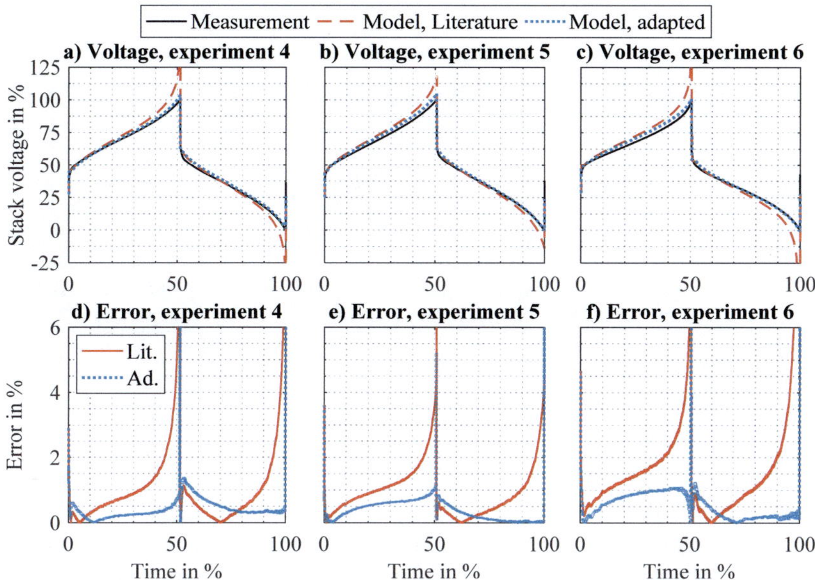

Figure 4-5: Validation with experimental data of manufacturer C with initial and adapted model, **individual scaling factor** for the manufacturer, experiments 4 – 6

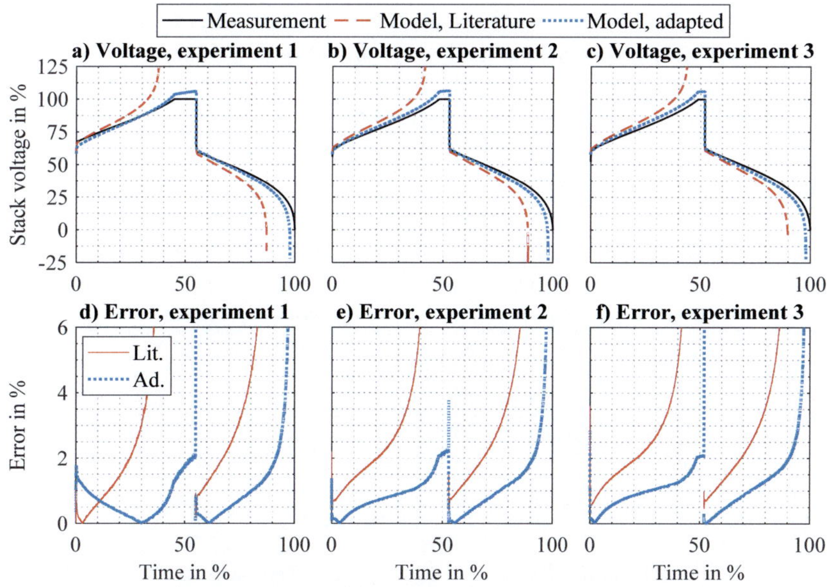

Figure 4-6: Validation with experimental data of manufacturer A with initial and adapted model, **average scaling factor of 2.38**

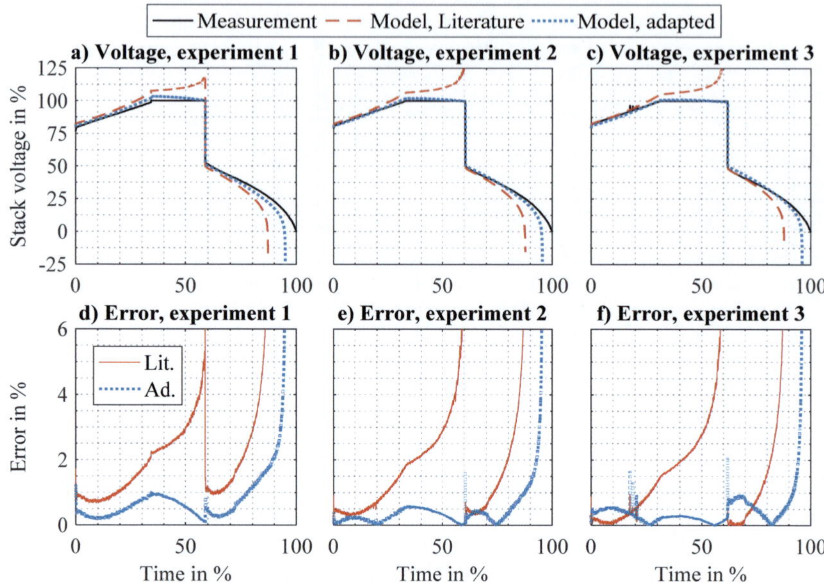

Figure 4-7: Validation with experimental data of manufacturer B,
average scaling factor of 2.38, experiments 1 – 3

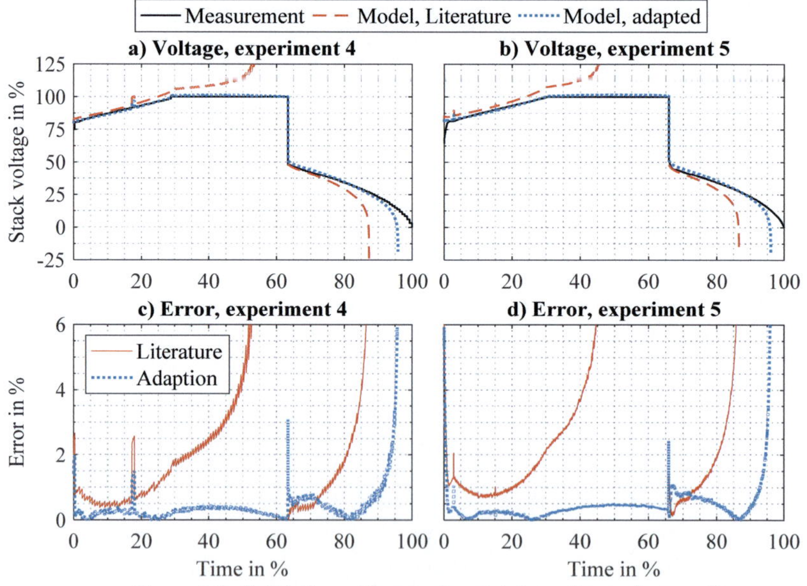

Figure 4-8: Validation with experimental data of manufacturer B,
average scaling factor of 2.38, experiments 4 and 5

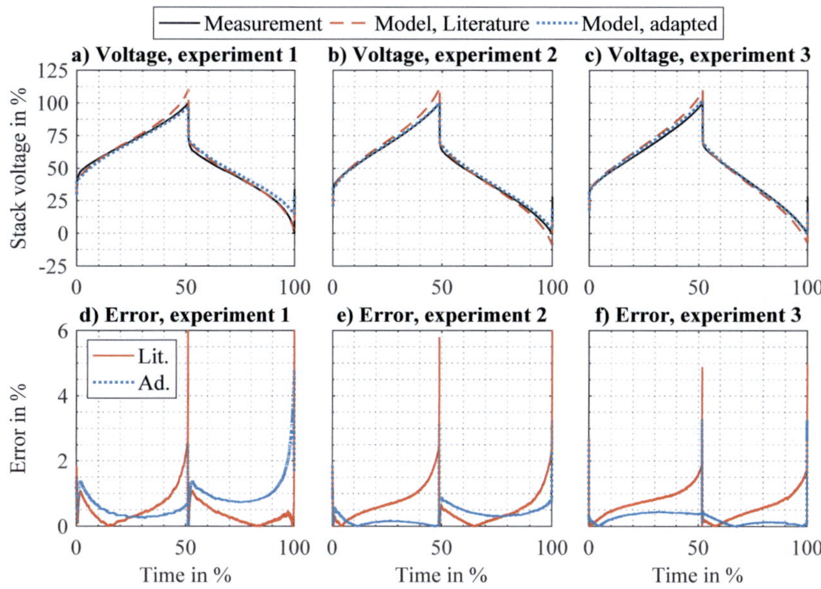

Figure 4-9: Validation with experimental data of manufacturer C, **average scaling factor of 2.38**, experiments 1 – 3

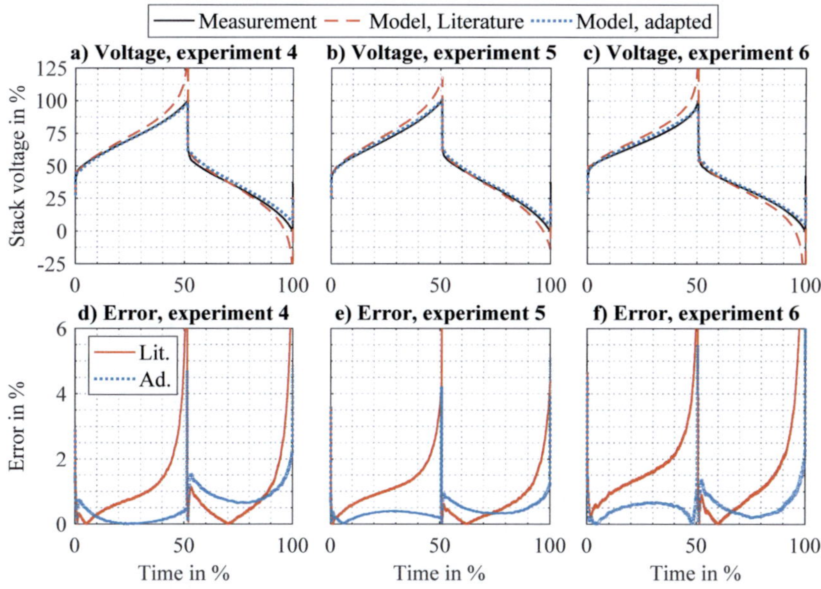

Figure 4-10: Validation with experimental data of manufacturer C, **average scaling factor of 2.38**, experiments 4 – 6

Chapter 5

Model-based cell design

5.1 Current state of science

While optimization of single-cell designs is common, cell designs are scarcely studied in stack context. Many design studies deal with replacing the simple flow-through electrode by more sophisticated designs [68, 70, 80–84]. Newly proposed designs comprise interdigitated or serpentine structures that significantly improve the battery performance. However, the manufacturing process of large-scale cells incorporating these structures is still very challenging. Hence, in this work, only simple flow-through electrodes are considered.

In [46], electrode height and compression are varied to identify a combination of a low area specific resistance (ASR) and an acceptable hydraulic resistance of the electrode. In this work, electrode height and compression is kept constant, to limit the number of studied design variations.

In [85], computational fluid dynamic (CFD) simulations are applied to optimize the internal electrolyte distribution of a commercial cell design. Similar to this work, the study also uses a meander-shaped channel. However, the modeled structures for electrolyte distribution are much more complex, which is beyond the scope of this work.

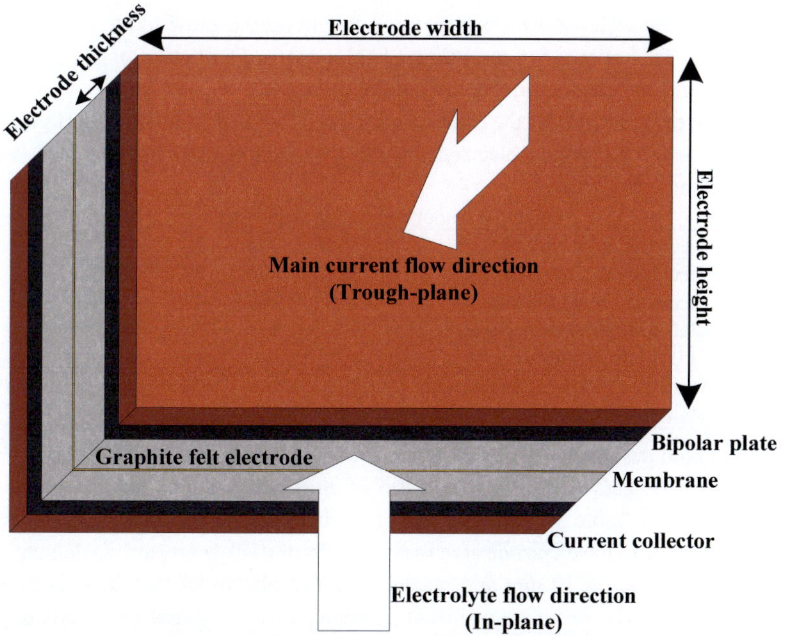

Figure 5-1: Simplified scheme of a redox flow cell

5.2 Initial considerations

5.2.1 Design parameters

In this work, the electrode area, the channel length and the channel width are subject to variation. The ratio of electrode height to width and the electrode thickness is not varied. The assignment of the denominations height, width and thickness is shown in Figure 5-1.

The ratio of electrode width to height is selected to be 1.5. Small ratios are not useful because narrow and height electrodes introduce two main disadvantages. First, the hydraulic resistance of a flow-through electrode obviously increases linearly with its height. Secondly, the concentration of vanadium ions in the electrolyte is going to face large variations as it passes through such an electrode. The only advantage of a narrow electrode is a homogenous fluid flow distribution across the whole cross-sectional area. Large width-to-height ratios lower the pressure drop in the electrode and obtain a more homogenous SoC distribution within the electrode.

However, very wide electrodes are challenging in terms of distributing the electrolyte flow homogeneously over the whole cross-sectional area. Although the high hydraulic resistance of the felt electrode intrinsically equalizes the electrolyte flow to a certain extent, additional measures to support the distribution are required for very wide electrodes. The design of special flow distribution structures within the cell is beyond the scope of this work.

Consequently, the selected electrode width-to-height ratio is close to the lower limit of useful values. This allows for assuming a homogenous distribution of the electrolyte flow across the total cross-sectional area of the electrode.

The second fixed parameter is the electrode thickness. In order to avoid a fourth variable design parameter, a constant electrode height of 4 mm is used in this work, which represents a moderate value.

Table 5-1 Input cell design parameter

Electrode area in cm²	1000	2000	3000	4000
Electrode width in mm	387	548	671	775
Electrode height in mm	258	365	447	516
Electrode thickness in mm	4	4	4	4
Cell thickness in mm	10	10	10	10
Manifold diameter in mm	30	40	50	60
Manifold geometry factor in m⁻¹	14.1	8.0	5.1	3.5

In the presented design study, electrode areas of 1,000 cm², 2,000 cm², 3,000 cm² and 4,000 cm² are studied. With the fixed ratio of electrode width to height, widths and heights shown in Table 5-1 are derived. The diameter of the internal manifold is increased along with the electrode area to account for the larger electrolyte demand. The overall cell thickness is 10 mm and results from two electrodes as well as membrane and bipolar plate. The resulting manifold geometry factor required for calculating the shunt currents is also given in Table 5-1.

5.2.2 Highest applied flow rate

Cell design is strongly affected by the highest applied flow rate. In practice, it is important to comply with maximum pressure limits to guarantee for a safe and leakage-free operation over the whole lifetime.

In terms of efficiency, optimal flow rate and cell design are cross-linked. A long and narrow channel leads to a lower optimal flow rate, due to the additional pressure drop. Thereby, it decreases the shunt current losses due to its higher ionic resistance. Energy required for pumping the electrolyte, losses due to shunt currents and losses due to concentration overpotential thus need to be considered holistically to achieve a system design with the highest possible efficiency.

However, the question is where to start. The flow rate cannot be optimized without a cell design and the cell design cannot be carried out without knowing the maximum applied flow rate. In this work, this circular relation is interrupted by determining a combination of a maximum tank SoC, a maximum cell voltage and a nominal current density during the charging process.

In practice, the useable SoC range of the battery is limited by the cell voltage, in particular if large current densities are applied. If the cell voltage gets too high, the aqueous electrolyte is electrolyzed and hydrogen and oxygen are going to evolve. Hydrogen evolution is a safety risk, lowers the Coulomb efficiency and imbalances the electrolyte.

To derive the required flow rate from SoC, upper cell voltage boundary and nominal current density, a simplified mathematical model is deployed on the basis of the Eqs. (2-43), (2-44) and (2-46) on page 35 and 36. For charging operation, the cell voltage is computed using Eq. (5-1) which is resolved in more detail in Eq. (5-2).

$$E_{C,charging} = E_{EMFC} + I_C R_C + E_{COPcharging-} + E_{COPcharging+} \qquad (5\text{-}1)$$

The equations (5-1) and (5-2) comprise four parts. The first part accounts for the cell EMF, calculated with the Nernst equation considering the actual cell SoC. In steady state and under negligence of all Coulomb losses, the cell SoC can be derived as shown in Eq. (5-3) [17]. The second part accounts for the ohmic overpotential. The third part accounts for the concentration overpotential of the negative half-cell and the fourth part for the concentration overpotential of the positive half-cell.

As the flow rate can be found in the logarithmic term of the OCV as well as in the concentration overpotential, Eq. (5-2) cannot be solved analytically. Hence, the required flow rate is derived numerically, as shown in Figure 5-2.

In this work, the maximum SoC to reach during a charging process with the nominal current density of 100 mAcm^{-2} while complying to an upper cell voltage limit of 1.65 V is chosen to be 80 %.

The crossings of the voltage curves with the horizontal line that indicates the upper cell voltage limit in Figure 5-2 sets the nominally required flow rate. For a flow battery, the fraction of applied over stoichiometric flow rate is often denoted as flow factor [30]. For more information regarding the flow rate, the reader is referred to the flow rate optimization section, starting on page 119.

$$E_{C} = \tilde{E}^{0} + 2\frac{GT}{F}\ln\left(\frac{SoC_{C}}{1\text{-}SoC_{C}}\right) + I_{C}R_{C}$$

$$-\frac{GT}{F}\ln\left(1 - \frac{I_{C}}{2.38A_{E}1.608\cdot10^{-4}F\left(Q_{C}/CSA_{E}\right)^{0.4}(1-SoC_{C})c_{V}}\right) \tag{5-2}$$

$$-\frac{GT}{F}\ln\left(1 - \frac{I_{C}}{2.38A_{E}2.613\cdot10^{-4}F\left(Q_{C}/CSA_{E}\right)^{0.4}(1-SoC_{C})c_{V}}\right)$$

$$SoC_{C} = SoC_{T} + \frac{I}{2Fc_{V}Q_{C}} \tag{5-3}$$

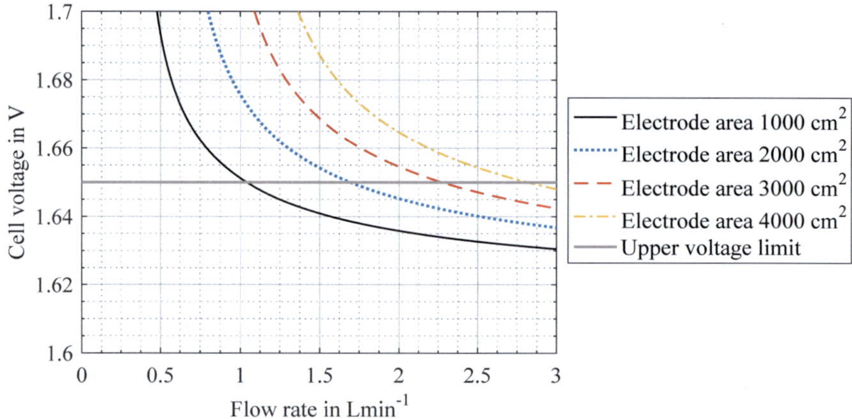

Figure 5-2: Cell voltage for charging operation with 100 mAcm^{-2} at 80 % SoC over applied flow rate

Table 5-2: Flow rates for charging operation of a 40-cell stack with a current density of 100 mAcm^{-2} at 80 % SoC

Electrode area	1000 cm²	2000 cm²	3000 cm²	4000 cm²
Stoichiometric flow rate	7.8 Lmin⁻¹	15.5 Lmin⁻¹	23.3 Lmin⁻¹	31.1 Lmin⁻¹
Flow rate for upper voltage limit	41.6 Lmin⁻¹	67.8 Lmin⁻¹	91.0 Lmin⁻¹	112.8 Lmin⁻¹
Flow factor	5.3	4.4	3.9	3.6

5.3 FEA simulations

With the electrode dimensions, according to Table 5-1, the design process using the FEA is carried out. Channel geometry is varied in length and width. Channel length is governed by the meander count. Between the two manifolds, as much space as possible is used to place the channel, but some space has to be reserved for manufacturing reasons and gaskets.

The basic channel geometry comprises one and a half meanders and is shown in Figure 5-3. An extra meander is added to the extended channel, as shown in Figure 5-4. For each of the four different electrode areas and each of the two meander counts, three different channel widths are studied. For each cell design, the geometry factor is derived as described in Eq. (2-35) on page 29. Total pressure drop across the cell (inlet channel, inlet funnel, graphite felt, outlet funnel and outlet channel) is evaluated for two different flow rates in order to derive the parameters according to Eq. (2-80) on page 51.

In total, twenty-four different designs are generated, as shown in Table 5-3. The presented approach yields a realistic set of input parameters for the design process. The geometry factor of the designs as well as the pressure drop at maximum applied flow rate are within reasonable ranges.

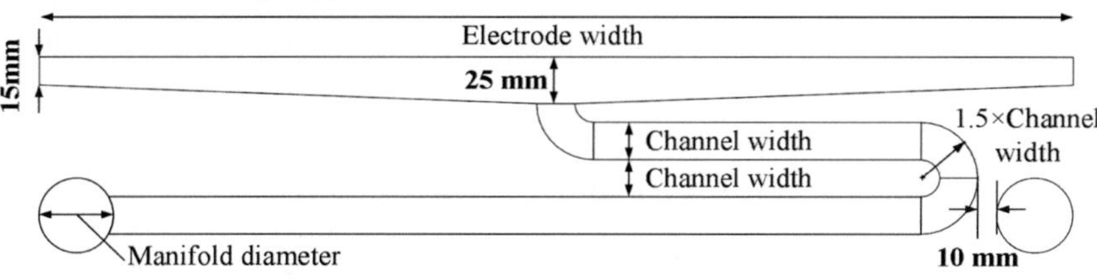

Figure 5-3: Channel design with 1.5 meanders

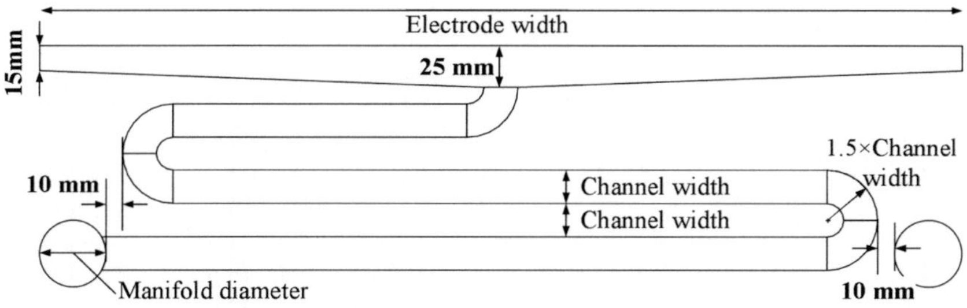

Figure 5-4: Channel design with 2.5 meanders

To facilitate presenting the results in the next sections, the different designs are indexed with a pair of numbers, separated by a decimal point. The first number refers to the electrode area. When multiplied by a factor of 1,000, it gives the electrode area of the respective design in cm^2. The second number refers to the channel design variation. The numbers 1, 2 and 3 denote designs with 1.5 meanders, whereas the numbers 4, 5, and 6 refer to designs with 2.5 meanders. The numbers 1, 2 and 3, as well as 4, 5 and 6, relate to an increasing geometry factor within the respective meander count.

The comparison of cell designs that comprise electrode areas between 1,000 cm^2 and 4,000 cm^2 is challenging. As shown in Table 5-2, flow rate requirements almost vary by a factor of three. Hence, for smaller electrode areas, finer channel structures can be studied. While a larger pressure drop not necessarily results in a lower system efficiency, it certainly increases the requirements for the manufacturing process, regarding sealing and handling of arising forces in operation.

Table 5-3: Geometry parameters and pressure drop at nominal flow rate of evaluated cell designs according to Table 5-2 on page 80

Design	Electrode area	Meander count	Channel width	Channel length	Geometry factor	Nominal pressure
1.1			10 mm	481 mm	16,036 m^{-1}	25.1 kPa
1.2		1.5	7.5 mm	479 mm	21,299 m^{-1}	28.0 kPa
1.3	1000 cm^2		5 mm	476 mm	31,735 m^{-1}	36.2 kPa
1.4			10 mm	786 mm	26,209 m^{-1}	28.4 kPa
1.5		2.5	7.5 mm	783 mm	34,828 m^{-1}	33.0 kPa
1.6			5 mm	779 mm	51,928 m^{-1}	49.9 kPa
2.1			20 mm	697 mm	11,644 m^{-1}	37.2 kPa
2.2		1.5	10 mm	689 mm	22,971 m^{-1}	45.7 kPa
2.3	2000 cm^2		7.5 mm	686 mm	30,474 m^{-1}	52.7 kPa
2.4			20 mm	1140 mm	19,006 m^{-1}	40.6 kPa
2.5		2.5	10 mm	1129 mm	37,629 m^{-1}	53.2 kPa
2.6			7.5 mm	1124 mm	49,963 m^{-1}	64.8 kPa
3.1			20 mm	851 mm	14,186 m^{-1}	51.3 kPa
3.2		1.5	15 mm	845 mm	18,788 m^{-1}	55.7 kPa
3.3	3000 cm^2		10 mm	839 mm	27,965 m^{-1}	65.5 kPa
3.4			20 mm	1404 mm	23,405 m^{-1}	57.4 kPa
3.5		2.5	15 mm	1385 mm	30,768 m^{-1}	63.6 kPa
3.6			10 mm	1374 mm	45,785 m^{-1}	78.8 kPa
4.1			20 mm	974 mm	16,227 m^{-1}	66.0 kPa
4.2		1.5	15 mm	968 mm	21,504 m^{-1}	71.9 kPa
4.3	4000 cm^2		10 mm	960 mm	31,985 m^{-1}	86.3 kPa
4.4			20 mm	1592 mm	26,525 m^{-1}	74.5 kPa
4.5		2.5	15 mm	1582 mm	35,145 m^{-1}	83.0 kPa
4.6			10 mm	1565 mm	52,159 m^{-1}	105.0 kPa

Table 5-4: Coefficients β and γ for calculating the non-linear hydraulic resistance of the channels according to Eq. (2-80) on page 51

Design	Electrode area	Meander count	Channel width	β in 10^7 Pa·s·m^{-3}	γ in 10^9 Pa·(s·m^3)^2
1.1			10 mm	3.41	2.98
1.2		1.5	7.5 mm	3.69	5.10
1.3	1000 cm^2		5 mm	4.43	11.56
1.4			10 mm	3.81	4.10
1.5		2.5	7.5 mm	4.27	7.14
1.6			5 mm	4.94	32.48
2.1			20 mm	3.20	0.86
2.2		1.5	10 mm	3.65	3.50
2.3	2000 cm^2		7.5 mm	4.09	5.08
2.4			20 mm	3.44	1.35
2.5		2.5	10 mm	4.26	3.98
2.6			7.5 mm	4.90	7.40
3.1			20 mm	3.26	0.84
3.2		1.5	15 mm	3.47	1.35
3.3	3000 cm^2		10 mm	3.89	2.85
3.4			20 mm	3.60	1.20
3.5		2.5	15 mm	3.91	1.88
3.6			10 mm	4.59	3.98
4.1			20 mm	3.36	0.83
4.2		1.5	15 mm	3.57	1.35
4.3	4000 cm^2		10 mm	4.05	2.88
4.4			20 mm	3.71	1.33
4.5		2.5	15 mm	4.09	1.75
4.6			10 mm	4.86	3.84

Chapter 6

Evaluation of the cell designs in a single-stack system

6.1 Current state of science

The main difference between considering a single-cell or considering several cells in a stack context are the shunt currents.

First works dealing with modeling and computing shunt currents date back to 1942 [34]. In 1976, the NASA (National Aeronautics and Space Administration) was the first to report a shunt current model for a redox flow battery [86].

The first model to incorporate the SoC dependency of the ionic electrolyte resistance is presented in [39]. A stack with up to 20 cells is investigated. The shunt current phenomenon in multi-stack strings is studied in [9, 40] with comparatively simple dynamic models, which for example do not consider the vanadium crossover. In [40], locations of electrolyte inputs and outputs are varied to study the effect on shunt currents. In [9], the author of this work examines a system comprising six 30-cell stacks. The stacks are operated as single-stacks, two-stack, three-stack and six-stack strings. Diameter of common hydraulic piping is found to affect shunt currents significantly. A comparable study is presented in [87]. Herein, system compactness is additionally taken into account. A total number of 120 cells is virtually assembled in up twelve stacks.

Variations in manifold radius and channel length are studied in [38] to reduce the effect of shunt currents on battery efficiency. A long and narrow channel is found to substantially reduce shunt currents. However, it is not evaluated if the increased hydraulic resistance of this channel still allows for a save operation in terms of pressure drop and an efficient operation in terms of pump power demand.

Another approach for reducing the negative impact of shunt currents on efficiency is to assemble short stacks that use a unipolar instead of a bipolar plate in between neighboring cells [88]. By means of intercalated current collectors, all cells of the short stack are electrically connected in parallel. While the authors in [88] claim an efficiency gain of up to 10 %-points, their approach causes a strongly increased effort in piping and stack assembling.

More recently, the lumped-parameter shunt current model has been replaced by an actual three-dimensional model [89]. However, the presented results are in accordance with the results from previously presented models.

6.2 System design

To evaluate the impact of different designs on round-trip system efficiency (RTSE) and accessible battery discharge capacity, a test system is set up for each design. Each system consists of one 40-cell stack with one pair of tanks. Tank volume is scaled accordingly to the total electrode area to obtain comparable charging and discharging times and conditions. For all studied designs, the ratio between energy and power rating is identical. For the 1000-cm², 2000-cm², 3000-cm² and 4000-cm² cell, the tank volumes are 250 L, 500 L, 750 L and 1,000 L each, respectively.

As the system only exists as a model, there is no exact piping plan available. Orifices and pipe lengths of the estimated external hydraulic circuit are the same for all designs. Per half-side, a total pipe length of 6 m is assumed. Usually, piping is mostly carried out using plastic pipes. This is primarily the case for connecting tanks and pumps. The connection of the stack is usually carried out using flexible tubes. However, pipes and tubes are identically modelled in this work. In single stack test systems, all pipes and tubes have the same diameter. This diameter is equal to stack manifold diameter. Therefore, no expansion or contraction, except for tank inlet and outlet, needs to be considered.

In terms of orifices, a total loss coefficient of 5.82 per half-side is assumed. This results from eight 90°-bends (k_L=0.3 each), tank inlet (k_L=1) and tank outlet (k_L=0.42). An additional loss coefficient of two is added per half-side to account for connection resistances and sensors (e.g., temperature and/or flow rate).

The pumps are assumed to have nominal capacities that comply with the flow rates according to Table 5-2 on page 80. Further, they have a lower limit of 10 % of nominal capacity, which cannot be undershoot as long as the pump runs. The flow rate dependent pump efficiency is shown in Figure 2-21 on page 55.

6.3 Evaluation methodology

6.3.1 Methodology

In the following, the terms round-trip-system-efficiency (RTSE) is used as a synonym for the system efficiency, η_{Sys}.

RTSE and discharging capacity of every design are determined using constant current cycles with four different current densities. The cycles are bounded by a combination of voltage and SoC limits. The charging process is finished when a cell voltage of 1.65 V or a tank SoC of 90 % is reached. The discharging process is finished when a cell voltage of 1.1 V or a tank SoC of 5 % is reached.

SoC limits are required for operation with low current densities, to compensate for imbalances in the operation of the individual cells. While for a high current density, overpotentials are high and system operation can be governed by cell voltage limits, it has to be governed by SoC limits for a low current density.

In practice, it is most likely that not all cells are equally well supplied with electrolyte. This can be caused by variations in the thickness of the graphite felt electrode, which leads to variations in the compression of the individual electrodes. Hence, the porosity and thus the permeability of the electrodes will vary to a certain extent, which directly influences the flow rate distribution on the individual cells. The cell with the least permeable electrode will suffer from the lowest flow rate.

If tank SoC is not limited during the charging process with a low current density, the SoC might reach values close to 100 % in the worst supplied cells, imposing these cells to the risk of overcharging. During discharging operation, a tank SoC of close to 0 % might occur. In this case, a polarity reversal can occur, if the discharging process is going on.

Prior to every cycle, the aforementioned pre-discharging process is carried out. Applied current densities correspond to 25 %, 50 %, 75 % and 100 % of the nominal current density, which is 100 mAcm^{-2}. A variable flow rate is applied, which is computed according to Faraday's first law of electrolysis scaled to the flow factor, see Section 8.5 on page 125. As the flow factor has a significant impact on RTSE and the optimal flow factor might be different for different designs, twelve different flow factors are studied for all cycles. Ten values are distributed evenly between a flow factor of 1.5 and the required flow factor according to Table 5-2. In addition, two larger flow factors are investigated. They are increased by one time and two times the equal step size between the ten values.

In any case, the flow factor for each design and each current density has to be large enough, to reach a tank SoC of 80 % during the charging process with the respective current density.

The RTSE of a particular design depends on four parameters: The electrode area, the channel design variation, the applied current density and the applied flow factor. To allow for a better comparison of the designs, an approach is presented that rates each of the twenty-four designs with one single value:

1. Calculate the flow factor that enables charging operation up to a tank SoC of 80 % for the studied current density.

2. Determine the flow factor that delivers the highest efficiency for a given design and a given current density by simulating round-trips.

3. If the most efficient flow factor is lower than calculated in step 1, replace it by the flow factor of step 1.

4. Calculate the current-weighted average efficiency as described in Eq. (6-1) out of the four studied current densities, simulated with the selected flow factors of step 3.

$$\emptyset RTSE = \frac{0.25\eta_{\mathrm{Sys}}(0.25i_{\mathrm{Nom}}) +0.5\eta_{\mathrm{Sys}}(0.5i_{\mathrm{Nom}})+0.75\eta_{\mathrm{Sys}}(0.75i_{\mathrm{Nom}})+\eta_{\mathrm{Sys}}(i_{\mathrm{Nom}})}{0.25+0.5+0.75+1} \qquad (6\text{-}1)$$

In Eq. (6-1), the four efficiency values are averaged by weighting each efficiency with the relative current density, referred to its nominal value. This implies that all current densities are equally often used in operation. For a single-stack system this assumption is reasonable, as the stack has to cover all power requirements from low partial to full load operation.

If all current densities are applied with identical frequencies, the efficiency for a higher current density is more important. This is because the absolute energy loss is proportional to the applied current density. The energy loss either has to be bought at the energy market in case the battery operates without a (renewable) energy source, or cannot be sold to the energy market, in case the battery operates together with a (renewable) energy source. As the efficiency is a relative value, it is weighted with the current density in order to account for the absolute losses.

6.3.2 Sample evaluation of the design 4.6

The presented stepwise evaluation methodology is applied to design 4.6 for illustration. RTSE values in dependence of applied current densities and a sample of applied flow factors are given in Table 6-1.

For design 4.6, a flow factor of 0.8, 1.1, 1.6 and 3.6 is required to reach a tank SoC of 80 % during a charging process with 25 mAcm^{-2}, 50 mAcm^{-2}, 75 mAcm^{-2} and 100 mAcm^{-2}, respectively. Note that flow factors below one are not applicable, as they correspond to a flow rate that is lower than required to fulfill Faraday's first law of electrolysis. This requirement is not considered in Eq. (5-2) on page 80.

Table 6-1: Derived RTSE values for design 4.6;
Bold: Highest RTSE for respective current density.
Gray box: RTSE of selected flow factor for respective current density.

Current density \ Flow factor	1.5	1.7	2.4	2.7	3.6	3.8	4.1
25 mAcm^{-2}	78.2 %	78.3 %	**78.6 %**	78.6 %	78.2 %	78.0 %	77.7 %
50 mAcm^{-2}	77.8 %	78.0 %	**78.2 %**	78.1 %	77.4 %	77.1 %	76.8 %
75 mAcm^{-2}	74.2 %	74.7 %	75.3 %	**75.4 %**	74.5 %	74.2 %	73.9 %
100 mAcm^{-2}	70.6 %	71.4 %	72.1 %	**72.4 %**	71.6 %	71.3 %	70.9 %

Table 6-2: Derived specific discharge capacity in WhL^{-1} for design 4.6;
Bold: Highest discharge capacity for respective current density.
Gray box: Discharge capacity of selected flow factor for respective current density.

Current density \ Flow factor	1.5	1.7	2.4	2.7	3.6	3.8	4.1
25 mAcm^{-2}	21.0	21.2	**21.2**	21.2	21.1	21.0	21.0
50 mAcm^{-2}	19.2	20.0	20.6	**20.6**	20.5	20.4	20.4
75 mAcm^{-2}	15.2	16.4	17.8	18.5	**18.6**	18.6	18.5
100 mAcm^{-2}	10.0	11.7	13.5	14.6	**15.2**	15.2	15.2

For the first three current densities, the identified most efficient flow factor is higher than the one, minimally required to reach a tank SoC of 80 %. Hence, the most efficient flow factor can be selected. For the nominal current density, the most efficient flow factor is 2.7. This flow factor does not allow for a charging operation up to a tank SoC of 80 % with the nominal current density. Therefore, it is replaced by a flow factor of 3.6. This increases the nominal capacity by 4.1 % while lowering the RTSE by 0.8 %-points.

Finally, from the four RTSE values in the gray boxes, a current-weighted average RTSE of 74.8 % can be derived for design 4.6.

6.4 Comparison of two sample designs

6.4.1 Dynamic simulations

To address the characteristics of different designs, two charging/discharging cycles are simulated with a current density of 25 mAcm^{-2} and 100 mAcm^{-2}, as shown in Figure 6-1 and Figure 6-2. Both cycles are conducted with cell design 1.1 (smallest electrode area, smallest channel geometry factor) and cell design 4.6 (largest electrode area, largest geometry factor). The flow factor according to the previously described selection process is applied to each case.

6.4.1.1 Lowest studied current density

During the cycles with the lowest current density, upper and lower cell voltage limits are not reached. Both, charging and discharging process are instead limited by tank SoC limits of 5 % and 90 %, as shown in Figure 6-1 a) and b).

The pumps of both systems supply their minimum flow rate for the predominant part of operation time, as shown in Figure 6-1 c). Nominal pump capacity, corresponding to the system's nominal flow rate according to Table 5-2 on page 80 is not exploited. The absolute value of the equivalent shunt current of design 1.1 is approximately threefold larger than of design 4.6, as shown in Figure 6-1 d). This is because of the short and wide channel of design 1.1. In addition, design 4.6 also has a fourfold increased electrode area and thus a fourfold higher current carrying capability. Hence, while up to 6.4 % of the externally applied current is lost for design 1.1, the highest share of lost current is only 0.6 % for design 4.6, as shown in Figure 6-1 e). The absolute value of the equivalent shunt currents follows the trend of the stack voltage. The threefold larger geometry factor and the fourfold larger electrode area results in a more than tenfold reduced shunt current sensitivity of cell design 4.6.

6.4.1.2 Nominal current density

During the cycles with nominal current density, charging and discharging process are governed exclusively by cell voltage limits, as shown in Figure 6-2 a) and b). The target tank SoC of 80 % at the end of the charging process, as postulated in Section 5.2.2, is reached by both designs.

Consequently, pump capacity is fully utilized, as shown in Figure 6-2 c). While it is only fully exploited at the very end of the charging process, it is fully utilized for a longer period of time during the discharging process. This is because the upper and lower voltage limits are not equidistant to the equilibrium voltage of 1.39 V. This results in a steeper voltage decrease at the end of the discharging process. Unsymmetrical cell voltage limits are used in practice because the VRFB is more tolerable to low cell voltages than to high cell voltages.

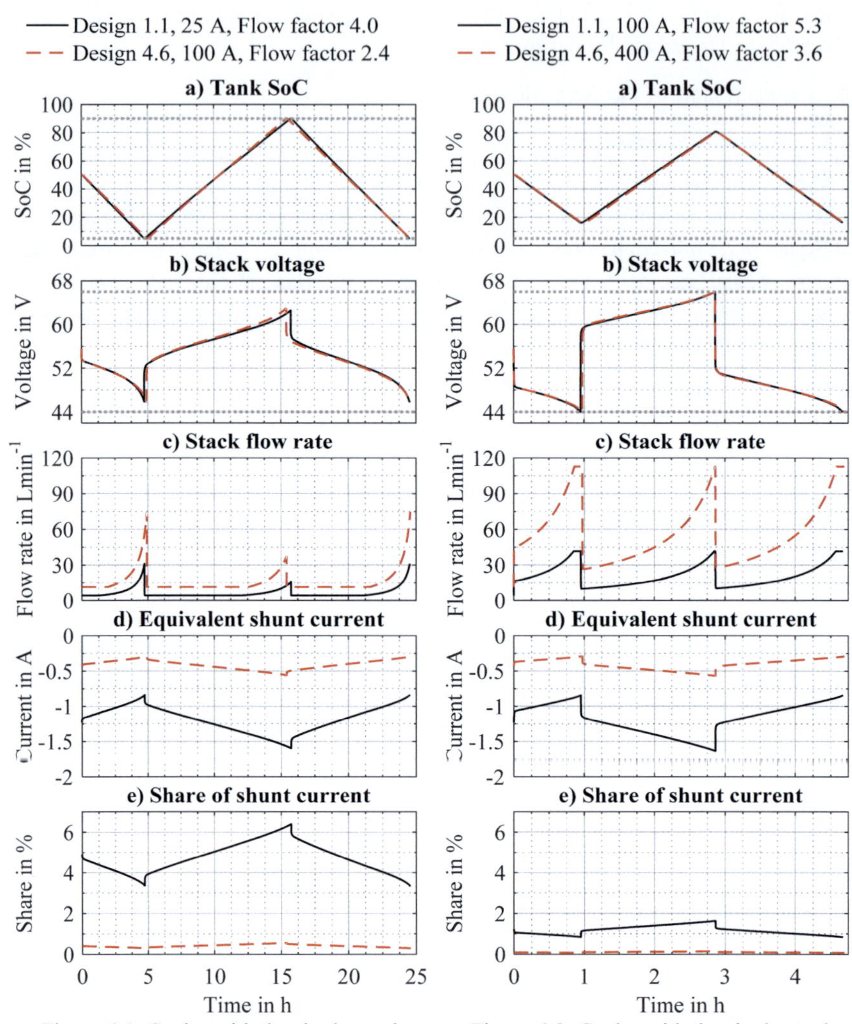

Figure 6-1: Cycles with the single-stack system with a current density of 25 mAcm^{-2}

Figure 6-2: Cycles with the single-stack system with a current density of 100 mAcm^{-2}

The equivalent shunt current evolves similar for both current densities, as shown in Figure 6-2 d). Obviously, the equivalent shunt current for operation with nominal current density hardly deviates from the equivalent shunt current for operation with the minimal studied current density. However, now that the batteries operate with their nominal current density, the impact of shunt currents decreases. Design 1.1 still loses up to 1.6 % of the externally applied current, whereas design 4.6 only loses a negligible share of less than 0.1 %, as shown in Figure 6-2 e).

6.4.1.3 Loss distribution

The losses accumulated during the previously simulated cycles can be related to ohmic losses, losses caused by vanadium crossover, losses due to concentration overpotential (COP), losses caused by shunt currents and pump energy demand.

In Figure 6-3, the losses are referred to the discharging capacity to allow for a comparison of different designs, although the deployed electrolyte volume varies by a factor of four.

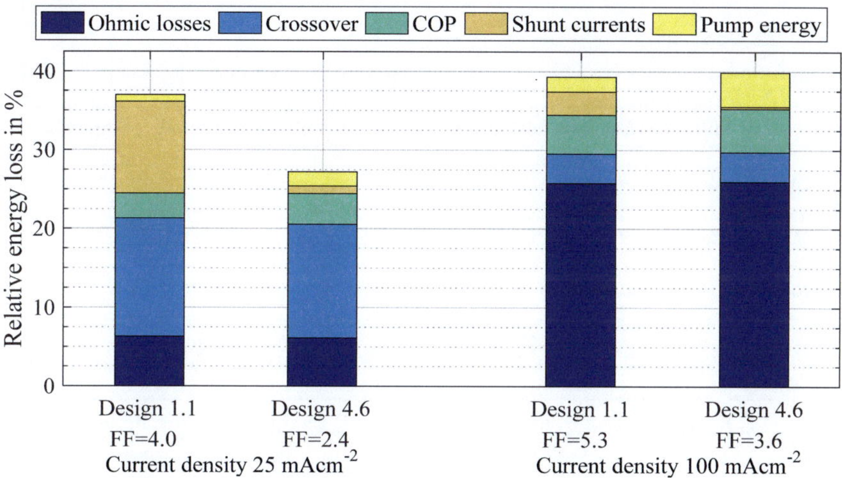

Figure 6-3: Relative energy loss of the designs 1.1 and 4.6 for a cycle with 25 mAcm^{-2} and 100 mAcm^{-2} (FF = Flow factor, COP = concentration overpotential)

For the operation at a low current density, the Coulomb losses, namely vanadium crossover and shunt currents account for the major loss share. The shunt current sensitive design 1.1 suffers significantly stronger from this parasitic process, compared to design 4.6. As expected, the latter shows a substantial reduction in shunt current losses.

For nominal current density, the losses related to overpotentials dominate. While ohmic losses are identical for both designs for a given current density, concentration overpotential shows some variations. The flow factor for design 1.1 is significantly larger than for design 4.6. Hence, the concentration overpotential losses of design 1.1 are lower than for design 4.6. However, although the deployed flow factor for design 4.6 is smaller, the required pump energy is larger. The shunt current losses of design 4.6 are negligible for operation at nominal current density.

6.4.2 Efficiencies of the sample designs in dependence of the flow factor

The designs 1.1 and 4.6 are studied in detail to illustrate the rating methodology of all designs. As shown in Figure 6-4 a) and b), the Coulomb efficiency hardly depends on the applied flow factor. The Coulomb efficiency increases with an increasing current density.

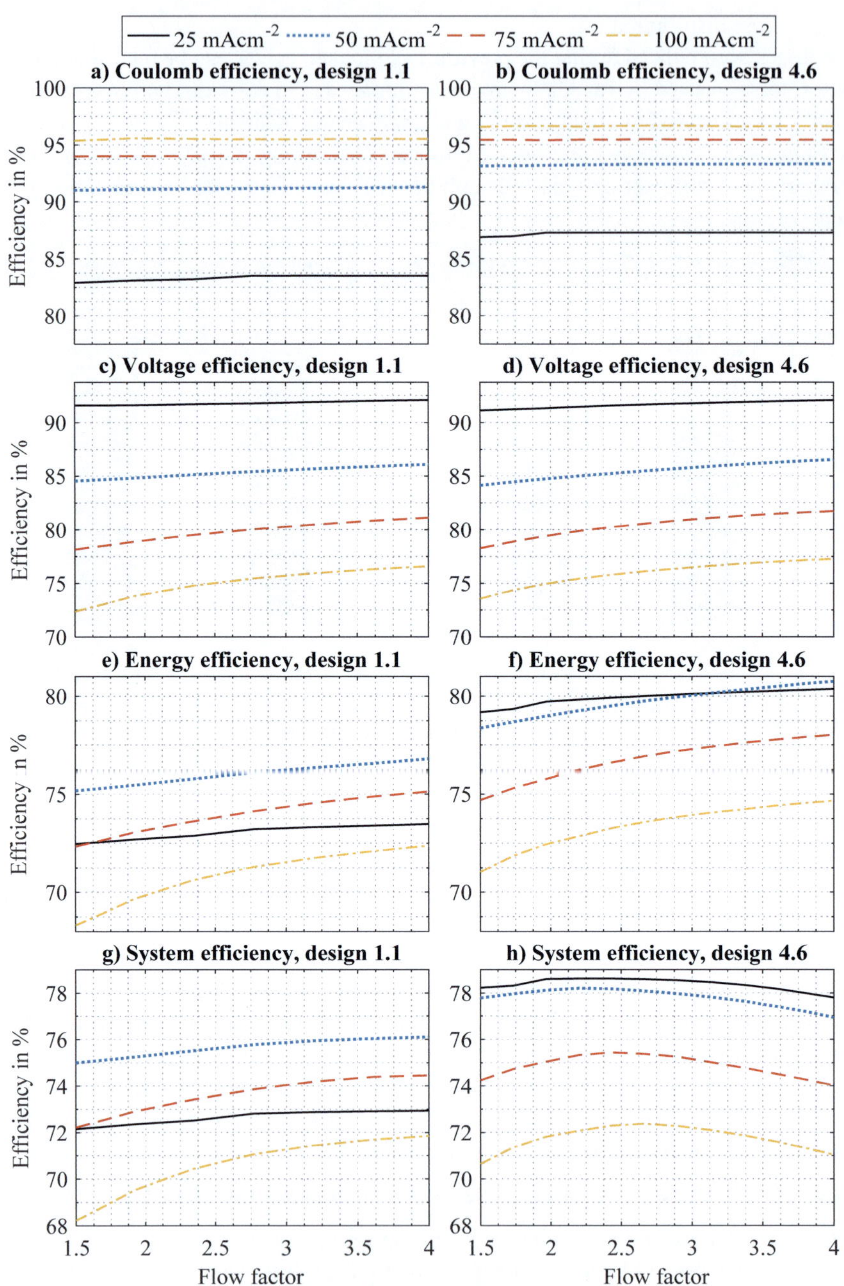

Figure 6-4: Coulomb, voltage, energy and system efficiency of design 1.1 and design 4.6 over the flow factor for different current densities

This is because a higher current density improves the ratio between desired charging and discharging currents and parasitic self-discharging currents due to vanadium crossover and shunt currents. However, it is noteworthy that for a given current density, the Coulomb efficiency of design 4.6 is substantially higher than the Coulomb efficiency of design 1.1. This is plausible for two reasons. First, design 4.6 has an approximately threefold larger channel geometry factor, which significantly reduces shunt currents, whereas vanadium crossover is not affected. Secondly, for a given current density, design 4.6 carries a fourfold increased charging/discharging current, because of its larger electrode area. The efficiency variation between the two designs is most significant for the lowest studied operation current density.

The voltage efficiency increases for both designs with an increasing flow factor, as shown in Figure 6-4 c) and d). The larger applied flow rate reduces the difference between cell and tank SoC and also reduces the concentration overpotential. The voltage efficiency significantly drops with an increasing current density due to the rising overpotentials.

The higher Coulomb and voltage efficiency of design 4.6 also increases the energy efficiency, as shown in Figure 6-4 f). As the Coulomb efficiency does not vary significantly with the flow factor, the energy efficiency follows the trend of voltage efficiency and increases with an increasing flow factor, as shown in Figure 6-4 e) and f). An increasing flow factor first increases both energy and system efficiency. This is mainly because concentration overpotential is lowered by the higher flow rate, while the additional pump power does not yet counterbalance the prevented overpotential losses. If the flow factor is increased beyond a certain value, pump power demand increases so strongly that the reduced concentration overpotential is overcompensated and the efficiency starts to decrease, as shown in Figure 6-4 g) and h).

For design 1.1, the efficiency peak is reached for a moderate current density of 50 mAcm^{-2}. For a lower current density, Coulomb losses lower the efficiency. For a higher current density, overpotentials and pump power increases strongly.

For design 4.6, the efficiency peak is already reached for 25 mAcm^{-2}, because of increased Coulomb efficiency due to reduced shunt currents.

In general, the required flow factors to obtain the efficiency peaks for all current densities are larger for the design 1.1 than for the design 4.6. This is further studied in Section 6.5.2 on page 96.

6.4.3   Discharge capacity of the sample designs

The discharge capacity mainly depends on two operational parameters, namely the applied current density and the applied flow rate or flow factor. If cell voltage limits are used to determine the end of the charging and discharging process, the discharge capacity declines with an increasing current density, as shown in Figure 6-5. Herein, the discharge capacity is referred to the total electrolyte volume in the stack and the tanks, to simplify the comparison between the two designs, whose stack and tank volumes differ by a factor of four. The decreasing discharge capacity can be explained by the higher stack voltage during the charging process with a high current density. Hence, the upper cell voltage limit is reached earlier and thus at a lower tank SoC, as shown in Figure 6-2 on page 90. During the discharging process, the higher current density lowers the stack voltage. Thus, the lower cell voltage limit is also reached earlier and thus at a higher SoC. A lower tank SoC at the end of the charging process and a higher tank SoC at the end of the discharging process automatically results in a lower discharge capacity. Hence, a larger current density leads to a lower discharge capacity. Regarding the flow rate, an increasing flow factor increases the discharge capacity, as long as the additionally used SoC limits of 5 % and 90 % are not reached during charging and discharging operations. A larger flow factor decreases the cell voltage during the charging process, as shown in Figure 5-2 on page 80 and thus increases the time until the upper cell voltage limit is reached. Consequently, the tank SoC at the end of the charging process is higher for a larger flow factor. During the discharging process, a larger flow factor increases the cell voltage which increases the time until the lower cell voltage limit is reached. Hence, more electric charge carriers can be released from the tank. Both, the higher tank SoC at the end of the charging process and thus at the beginning of the discharging process and the possibility to withdraw more electric charge carriers from the tank increases the discharge capacity.

However, in practice, the boost of discharge capacity by a larger flow factor is limited by the pump capacity. The discharge capacity boost obtained with lower current densities is additionally limited by the deployed SoC limits. These limits are the reason why the discharge capacity for a current density of 25 mAcm^{-2} is almost the same as for a current density of 50 mAcm^{-2}, as shown in Figure 6-5.

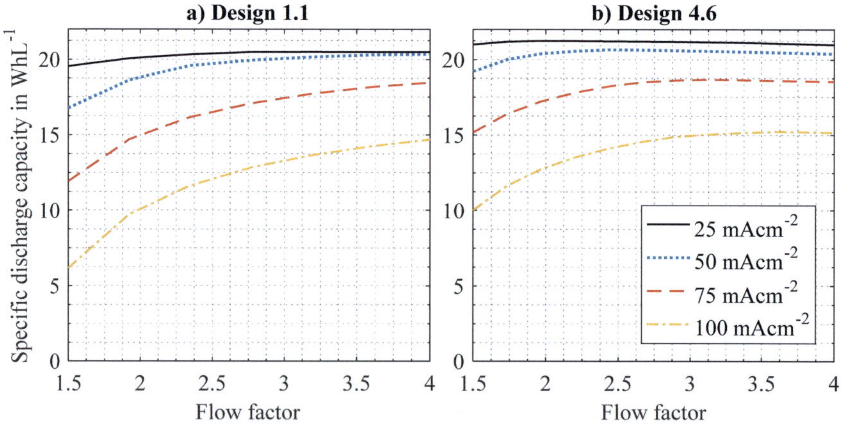

Figure 6-5: Specific discharge capacity over flow factor in dependence of current density for the designs 1.1 and 4.6

Design 4.6 yields a peak discharge capacity of 21.2 WhL^{-1} for a current density of 25 mAcm^{-2} exploiting 70.8 % of the theoretical specific capacity. The theoretical capacity is 30 WhL^{-1} for a total vanadium concentration of 1.6 molL^{-1}. At nominal current density, design 4.6 yields a specific discharge capacity of 15.2 WhL^{-1}, corresponding to an electrolyte exploitation of only 50.5 %.

Design 1.1 requires a higher flow factor to obtain a comparable discharge capacity for a particular current density. This effect is extensively studied in Section 6.5.2 on page 96.

6.5 Evaluation of all twenty-four cell designs

6.5.1 Coulomb efficiency

The Coulomb efficiency of all designs is compared for the lowest studied current density, which is 25 mAcm^{-2}. As shown in Figure 6-6, the Coulomb efficiency increases with an increasing channel geometry factor and an increasing electrode area. Between design 1.1, which has the lowest Coulomb efficiency, and design 4.6, which has the highest Coulomb efficiency, there is a gap of 7.5 %-points.

Increasing the active electrode area is found to be effective to increase the Coulomb efficiency. In fact, the Coulomb efficiency of the largest electrode with a short and wide channel (design 4.1) is higher than the Coulomb efficiency of the smallest electrode with a long and narrow channel (design 1.6).

For a channel configuration with a higher geometry factor, the electrode enlargement is less effective. In terms of Coulomb efficiency, both, electrode enlargement and narrower and longer channels are approximately equally effective. However, boosting the Coulomb efficiency by both measures, the electrode enlargement and a larger channel geometry factor, face a saturation effect.

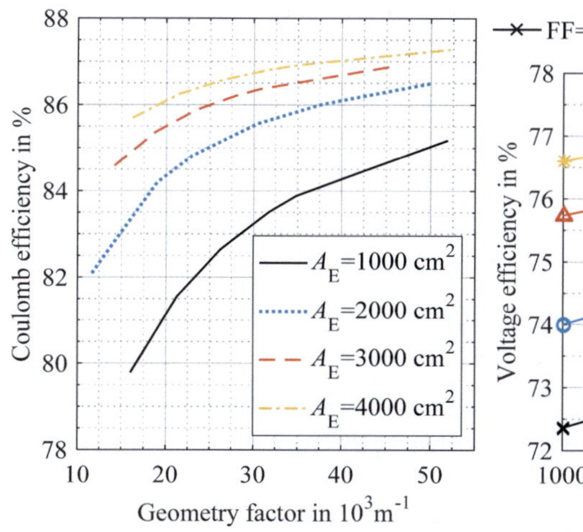
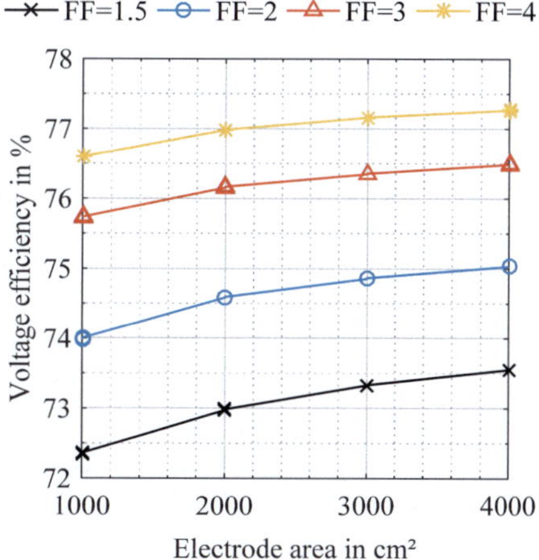

Figure 6-6: Coulomb efficiency over channel geometry factor and electrode area at 25 mAcm^{-2}

Figure 6-7: Voltage efficiency over electrode area and flow factor (FF) at 100 mAcm^{-2}

6.5.2 Voltage efficiency

The voltage efficiency of all designs is compared for the highest applied current density, which is 100 mAcm^{-2}. As shown in Figure 6-7, the voltage efficiency increases with electrode area and flow factor. While the latter is obvious, the positive effect of an enlarged electrode has to be further studied. As the specific ohmic resistance is identical for all electrode areas, the concentration overpotential is suspected to cause the phenomenon.

If the flow rate is controlled using the instantaneously derived stoichiometric requirements according to Faraday's first law of electrolysis, scaled by a given flow factor, the flow rate is proportional to the nominal current, as shown in Eq. (8-6) on page 125. Hence, we yield Eq. (6-2)

$$Q_C \propto I_C \tag{6-2}$$

The nominal current density is the same for all designs, as shown in Eq. (6-3).

$$i_C = \frac{I_C}{A_E} = \text{const.} \tag{6-3}$$

If we combine the Eqs. (6-2) and (6-3), we find that under the given boundaries, the applied flow rate is proportional to the electrode area, as shown in Eq. (6-4).

$$Q_C \propto I_C = i_C A_E \Leftrightarrow Q_C \propto A_E \tag{6-4}$$

Further, the cross-sectional area of the electrode is the product of electrode thickness, δ_E, and electrode width, w_E. For a fixes aspect ratio between electrode width and height, as it is the case in this work, the width can be computed from the electrode area and the aspect ratio. In fact, the cross-sectional area in fluid flow direction is then proportional to the square root of the electrode area, as shown in Eq. (6-5).

$$CSA_E = \delta_E w_E = \delta_E \sqrt{1.5 A_E} \Leftrightarrow CSA_E \propto \sqrt{A_E} \tag{6-5}$$

For the concentration overpotential, the electrolyte velocity in the electrode, v_{El}, is important. The velocity is equal to volumetric flow rate over cross-sectional area of the electrode in fluid flow direction, CSA_E, as shown in Eq. (6-6).

$$v_{El} = \frac{Q_C}{CSA_E} \tag{6-6}$$

If we combine the Eqs. (6-5) and (6-6), we find the electrolyte velocity to be proportional to the square root of the electrode area, as shown in Eq. (6-7). Hence, if the flow rate is adapted to applied current and tank SoC as shown in Eq. (8-6) on page 125, the electrolyte velocity increases with the electrode area, which is counter-intuitive.

$$v_{El} = \frac{Q_C}{CSA_E} \propto \frac{A_E}{\sqrt{A_E}} = \sqrt{A_E} \tag{6-7}$$

The relation, presented in Eq. (6-7) is illustrated in Figure 6-8 for the stoichiometric flow rate calculated from a tank SoC of 80 % and a current density of 100 mAcm^{-2}, multiplied by different flow factors. For a given flow factor, the fluid velocity increases with the electrode area. If the electrode area is increased by a factor of four, the fluid velocity doubles for a given flow factor, as shown in Figure 6-8 a). If we want to obtain a certain fluid velocity, we have to apply a larger flow factor for a smaller electrode area, as shown in Figure 6-8 b).

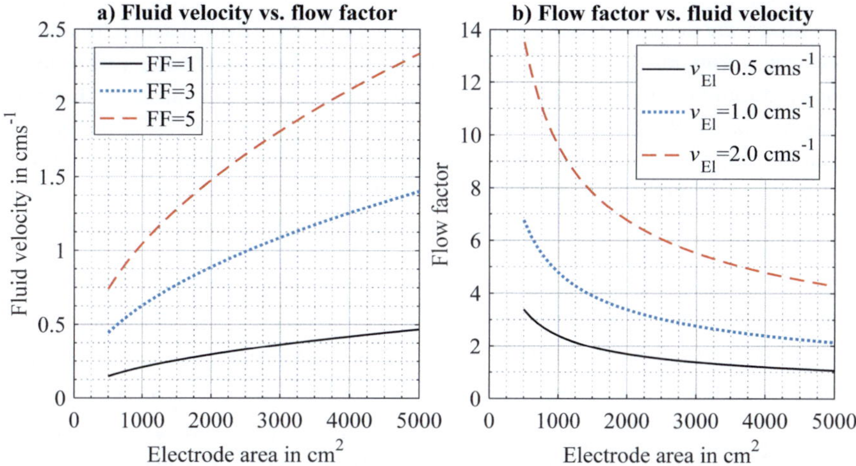

Figure 6-8: Impact of electrode area on fluid velocity for a given flow factor and vice versa

The fluid velocity directly affects the mass transfer coefficient, as shown in Eq. (2-70) on page 44. For a given flow factor, a larger electrode yields a higher fluid velocity. Hence, the mass transfer coefficient is increased, the concentration overpotential is decreased and the voltage efficiency rises. Or to rephrase: If a particular mass transfer coefficient should be reached to limit the concentration overpotential, a larger electrode requires a lower flow factor.

The dependence of the fluid velocity on the electrode area explains the increasing voltage efficiency of a larger electrode for a given flow factor, as shown in Figure 6-7. Note, that these considerations only apply, if we compare electrodes with different areas, but identical aspect ratios.

However, the flow factor affects the voltage efficiency significantly stronger than the electrode area. Hence, we can compensate the negative effect of a smaller electrode area on the voltage efficiency by applying a larger flow factor. Naturally, this comes at the costs of an increased pump power demand.

The holistic considerations in this design study reveal that the smallest electrode area can be operated as efficiently as the largest electrode area, if all loss mechanisms are taken into account, as shown in Table 6-6 on page 103. Nevertheless, the afore-mentioned phenomenon is present in many facets in this work. Amongst others, it also explains why smaller electrodes require a higher flow factor to operate most efficiently, as shown in the next section.

6.5.3 Optimal flow factor

In accordance with the just given explanation, the flow factors which yield the highest RTSE decrease with an increasing electrode area, as shown in Table 6-3. The smallest studied electrode requires a significantly larger flow factor for an optimal operation than the largest one.

It is noteworthy that a narrower and longer channel only has little impact on the optimal flow factor. Obviously, the higher pressure drop and thus the higher pump power demand does not introduce enough power loss to choose a significantly smaller flow factor and hence to accept higher concentration overpotential losses. For a particular electrode area, the optimal flow factor for different channel geometries only varies by 26 %. This is surprising, given that the channel geometries vary substantially. For a particular electrode area, the nominal pressure drop for the studied channel geometries varies by up to the factor of two, as shown in Table 5-3 on page 82.

Table 6-3: Deployed flow factors for all twenty-four designs and the lowest and the highest studied current density; In brackets: Optimal flow factor in terms of RTSE, if the applied flow factor deviates from the optimal flow factor

Area variation	Electrode area	Channel design variation					
		1 short wide	2 short medium	3 short narrow	4 long wide	5 long medium	6 long narrow
Current density 25 mAcm^{-2}							
1	1000 cm²	4.0	4.0	4.0	4.0	4.0	3.6
2	2000 cm²	3.4	3.1	3.1	3.4	3.1	2.8
3	3000 cm²	3.1	3.1	2.8	3.1	2.8	2.6
4	4000 cm²	2.7	2.7	2.7	2.7	2.4	2.4
Current density 100 mAcm^{-2}							
1	1000 cm²	5.3 (4.5)	5.3 (4.5)	5.3 (4.5)	5.3 (4.0)	5.3 (4.0)	5.3 (4.0)
2	2000 cm²	4.4 (3.8)	4.4 (3.4)	4.4 (3.4)	4.4 (3.4)	4.4 (3.4)	4.4 (3.4)
3	3000 cm²	3.9 (3.1)	3.9 (3.1)	3.9 (3.1)	3.9 (3.4)	3.9 (3.1)	3.9 (2.8)
4	4000 cm²	3.6 (3.1)	3.6 (2.9)	3.6 (2.7)	3.6 (2.9)	3.6 (2.7)	3.6 (2.7)

For the nominal current density, the optimal flow factor in terms of efficiency does not allow for a charging operation up to a tank SoC of 80 %. Hence, it is replaced by the flow factor that enables this operation. In this case, the optimal flow factor in terms of RTSE can be found in brackets, as shown in Table 6-3. For all studied lower current densities, the optimal flow factor in terms of RTSE is larger than the required flow factor to reach a tank SoC of 80 % during a charging process with the respective current density. The minimally required flow factor only depends on the electrode area, not on the channel design.

6.5.4 Auxiliary losses

The pump power demand is the only considered source of external or auxiliary losses in this work. Its influence on the RTSE is evaluated by studying the auxiliary efficiency, η_{Aux}. From Eq. (3-4) on page 59 we derive Eq. (6-8).

$$\eta_{Aux} = \frac{\eta_{Sys}}{\eta_{Ene}} \tag{6-8}$$

In general, the auxiliary efficiency is higher for a short and wide channel than for a long and narrow channel, as shown in Table 6-4. This is reasonable, as the latter introduces an additional hydraulic resistance and thus an additional pump power demand. However, what is more interesting is the fact that the auxiliary efficiency also decreases gradually for larger electrode areas.

Table 6-4: Auxiliary efficiency due to pump power demand of all twenty-four designs for a cycle with nominal current density and flow factors according to Table 6-3.

Area variation	Electrode area	Channel design variation					
		1 short wide	2 short medium	3 short narrow	4 long wide	5 long medium	6 long narrow
1	1000 cm²	98.3 %	98.2 %	97.8 %	98.2 %	97.9 %	97.2 %
2	2000 cm²	98.1 %	97.7 %	97.4 %	97.9 %	97.4 %	96.9 %
3	3000 cm²	97.8 %	97.7 %	97.3 %	97.6 %	97.4 %	96.8 %
4	4000 cm²	97.5 %	97.3 %	96.9 %	97.2 %	96.9 %	96.2 %

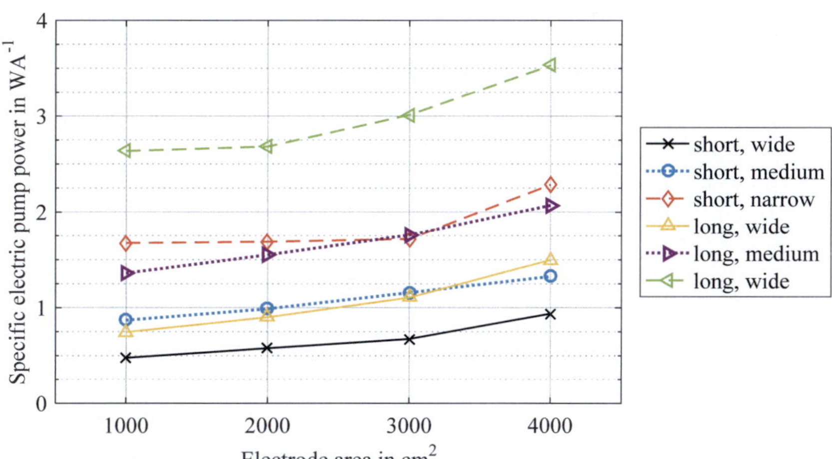

Figure 6-9: Electric pump power of all twenty-four designs referred to their nominal current for nominal flow rates according to Table 5-2 on page 80 to overcome the channel hydraulic resistance

This can be explained by examining the pump power, which is required to deliver the nominal flow rate according to Table 5-2 on page 80. In Figure 6-9, the pump power is referred to the nominal current of each electrode area. Although for each electrode area, the power varies with the channel design variation, there is a clear trend that a larger electrode area requires a larger specific pump power. The hydraulic resistance of the electrode itself is identical for all designs. This is because the ratio of electrode width and height is fixed, which results in a constant hydraulic resistance, as shown in Eq. (2-78) on page 49. Hence, this phenomenon is caused exclusively by the channel. Obviously for a larger cell, the pump power for the larger flow rate to overcome the hydraulic resistance of the channel increases more strongly than the useable electric power of the battery. It would be possible to choose larger channel widths or shorter channel lengths for the larger electrodes to reduce the hydraulic resistance. However, this also lowers the geometry factor and gives rise to larger shunt currents.

6.5.5 Shunt currents versus pump power

In a single-stack with a given number of cells, the equivalent shunt current is exclusively related to the channel geometry factor, as shown in Figure 6-10 for the studied 40-cell stack. The correlation between channel geometry factor and equivalent shunt current is precisely described by the hyperbola given in Eq. (6-9). The shunt current reduction by a larger geometry factor faces a strong saturation effect. Thus, beyond a geometry factor of 60,000 m^{-1}, the absolute value of the equivalent shunt current only declines slowly.

$$I_{\text{Shunt}}(Geo_{\text{Ch}}) = -20.96 \cdot 10^3 \, \text{Am}^{-1}/Geo_{\text{Ch}} \qquad (6\text{-}9)$$

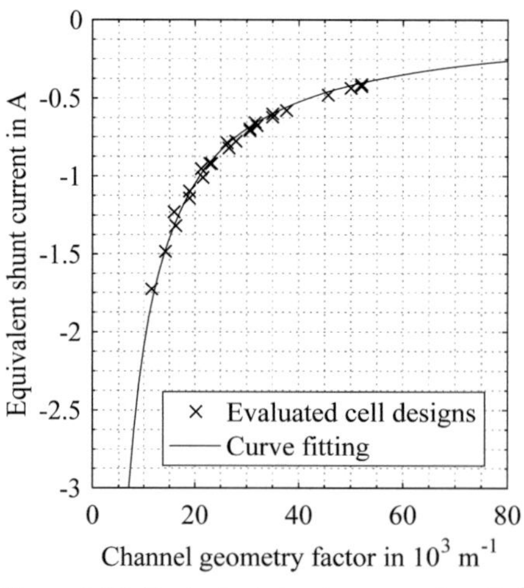
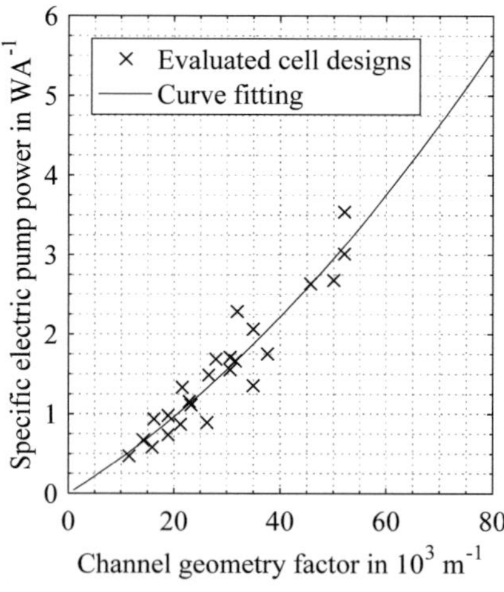

Figure 6-10: Equivalent shunt currents for all designs in a 40-cell stack for a tank SoC of 50 % and no load conditions.

Figure 6-11: Specific electric pump power referred to the electrode area for all designs in a 40-cell stack and the nominal flow rates according to Table 5-2 on page 80

The advantage of a large geometry factor in terms of shunt currents is directly opposed to the pump power, required to overcome the hydraulic resistance of the channel that generates the large geometry factor. Unfortunately, the pump power increases strongly with a higher geometry factor, as shown in Figure 6-11. Herein, the pump power to overcome the pressure drop of both, the incoming and outgoing channel is shown. It is referred to the nominal current of the electrode, to allow for a comparison of all designs. The applied flow rates correspond to the nominal flow rates, as shown in Table 5-2 on page 80.

However, while for the shunt currents, it is not important how the geometry factor is composed out of channel cross-sectional area and channel length, the pressure drop and thus the pump power depends on these two parameters. Nevertheless, a rule of thumb for how pump power and channel geometry are related can be provided, as shown in Eq. (6-10).

$$\frac{P_{\mathrm{PumpsCh}}}{I_{\mathrm{CNom}}}(Geo_{\mathrm{Ch}}) = 4.1{\cdot}10^{-5}\,\frac{\mathrm{Wm}}{\mathrm{A}}\,Geo_{\mathrm{Ch}} + 3.6{\cdot}10^{-10}\,\frac{\mathrm{Wm}^2}{\mathrm{A}}\,Geo_{\mathrm{Ch}}{}^2 \qquad (6\text{-}10)$$

The two relations (6-9) and (6-10) allow for estimated calculations of cell designs which are not explicitly studied in this work, e.g. an electrode with an area of 1,500 cm^2 and a nominal current of 150 A. If this electrode is equipped with a channel geometry factor of 30,000 m^{-1}, the additionally required electric pump power to overcome the hydraulic resistance of the channel can be estimated to be 233 W for a 40-cell stack. If the geometry factor is doubled, the power rises to 563 W.

At the same time, the absolute value of the equivalent shunt current is reduced by 50 % from 0.7 A to 0.35 A, as shown in Eq. (6-9). Multiplied with the equilibrium stack voltage of a 40-cell stack, this leads to a reduction in shunt current losses by 20 W. Hence, the additionally required electric pump power of 330 W for the total stack only reduces the shunt current losses by 20 W. However, if a variable flow rate is applied, it has to be taken into account that the nominal flow rate is only applied for a short period of time, while the shunt currents cause constant losses. In this case, an analytic assessment whether the additionally deployed pump power is rewarded with sufficient savings in shunt current losses is not possible.

6.5.6 Final evaluation using the round-trip system efficiency

6.5.6.1 Lowest studied current density

In general, cell design affects RTSE significantly stronger for a low current density, as shown in Table 6-5 for a current density of 25 mAcm^{-2}. The difference between lowest and highest efficiency is 5.7 %-points. In general, the findings regarding the RTSE for the battery operation with a low current density can be summarized as follows:

- The efficiency is higher for a narrow than for a wide channel.
- The efficiency is higher for a large than for a small electrode.
- The efficiency is higher for a long channel than for a short channel if comparable channel widths are deployed.

Table 6-5: RTSE values for two different current densities and all twenty-four designs

Area variation	Electrode area	Channel design variation					
		1 short wide	2 short medium	3 short narrow	4 long wide	5 long medium	6 long narrow
Current density 25 mAcm^{-2}							
1	1000 cm²	72.9 %	74.5 %	76.2 %	75.5 %	76.6 %	77.6 %
2	2000 cm²	74.8 %	77.2 %	77.8 %	76.7 %	78.2 %	78.5 %
3	3000 cm²	76.8 %	77.5 %	78.1 %	77.9 %	78.3 %	78.6 %
4	4000 cm²	77.6 %	78.0 %	78.4 %	78.3 %	78.5 %	**78.6 %**
Current density 100 mAcm^{-2}							
1	1000 cm²	71.8 %	72.1 %	72.2 %	72.3 %	72.4 %	72.2 %
2	2000 cm²	72.1 %	72.4 %	72.3 %	72.4 %	72.4 %	72.1 %
3	3000 cm²	72.3 %	72.3 %	72.2 %	**72.4 %**	72.3 %	72.0 %
4	4000 cm²	72.2 %	72.2 %	72.0 %	72.2 %	72.0 %	71.6 %

Consequentially, design 4.6 with the largest studied electrode and the largest studied channel geometry factor yields the highest RTSE.

6.5.6.2 Nominal current density

For cycles with nominal current density, variations in RTSE caused by different cell designs are significantly smaller. Lowest and highest RTSE are only 0.6 %-points apart. Neither the largest electrode nor the largest channel geometry factor yield the highest efficiency. For an electrode area of 1000 and 2000 cm², the shortest and widest channel yields the lowest efficiency. Hence, the negative effect of the shunt currents is still considerable for these cells, even for the operation with the nominal current density.

6.5.6.3 Current-weighted average RTSE

So far, different designs yield the highest RTSE for different current densities. Hence, the efficiencies are now weighted according to the applied current density, in order to identify the cell design which offers the best performance over the whole operation range.

It is an intrinsic property of the deployed rating procedure that the RTSE at nominal current density has the highest impact and the RTSE at lowest studied current density has the lowest impact, as shown in Eq. (6-1) on page 87. The variation in the current-weighted RTSE for the twenty-four designs is 1.6 %-points. The design that yields the highest current-weighted RTSE is neither the design that yields the highest RTSE for the lowest studied current density, nor the design that yields the highest RTSE for the nominal current density.

Cell design 2.5 complies best with the requirement of an efficient battery operation with variable current densities. It comprises a 2000-cm² electrode and a channel with a length of 1,129 mm, a width of 10 mm, a resulting geometry factor of 37,629 m^{-1}, a nominal flow rate of 67.8 Lmin^{-1} and a nominal pressure drop across the stack of 53.2 kPa.

Table 6-6: Current-weighted average RTSE values for all twenty-four designs

Area variation	Electrode area	Channel variation					
		1 short wide	2 short medium	3 short narrow	4 long wide	5 long medium	6 long narrow
1	1000 cm²	73.6 %	74.2 %	74.7 %	74.5 %	74.9 %	75.0 %
2	2000 cm²	74.2 %	75.0 %	75.1 %	74.8 %	**75.2 %**	75.1 %
3	3000 cm²	74.8 %	74.9 %	75.0 %	75.1 %	75.1 %	75.0 %
4	4000 cm²	74.9 %	75.0 %	74.9 %	75.0 %	75.0 %	74.8 %

6.6 Conclusion

The single-stack design study reveals three important correlations:

1. Enlarging the electrode and increasing the channel geometry factor using a long and narrow channel is equally effective to limit the impact of shunt currents.
2. For a constant aspect ratio, a larger electrode requires a lower flow rate to obtain the same fluid velocity and thus the same mass transfer coefficient as a smaller electrode. Hence, referred to the electrode area, the optimal flow rate for a smaller electrode is larger than for a larger electrode.
3. The large electrolyte demand of a large electrode slightly lowers the auxiliary efficiency. This means that the relative pump power demand tends to increase.

However, if the presented design methodology is obeyed, electrode areas between 1,000 and 4,000 cm^2 can be used to construct almost equally efficient VRFBs. The key is to equip the electrode with a channel design adapted to its area and to identify optimal flow factors. In this case, the different loss mechanisms, namely shunt currents, concentration overpotential and pump power demand can be balanced in a way that all electrode areas obtain comparable efficiencies. This fact promotes an excellent scalability of the redox flow technology.

The presented approach includes a simple flow rate optimization by identifying the optimal flow factors for each design and each current density. It is inevitable to include a flow rate optimization into the design study. If all designs are operated with the same flow factor, smaller electrode areas would suffer from the aforementioned phenomenon of lower fluid velocities. Hence, in this study, larger optimal flow factors are identified and deployed for these designs. The additional pump power demand of the larger flow factors is tolerable, due to the lower pressure drop across the channels of the cells with a smaller electrode.

Consequently, the smaller electrode areas are surprisingly competitive, compared to the larger electrodes, given their advantages in terms of both, shunt currents and concentration overpotential.

For a single-stack system, which uses all reasonable power levels equally frequent, a long but medium wide channel in combination with an electrode area of 2000 cm^2 is the best choice. However, as the total difference between all studied designs is rather small, additional parameters in terms of costs and manufacturing can be considered to identify the most suitable design. It is most likely that the small difference in efficiency will be outweighed by these additional factors. If lower power levels are used significantly more frequently than high-power levels, cell design gains significance. In this case, the largest possible electrode area with a long and narrow channel will represent the most efficient design.

A single-stack system most likely operates as a stand-alone system, e.g., for the supply of households, farms or telecommunication towers. Hence, it is important to adapt power and energy rating of the single unit according to customer's requirements. While the energy rating is easily scalable by deploying more or less electrolyte, the power rating can be varied by using different electrode areas. Thus, it is good to see that electrode areas between 1,000 and 4,000 cm^2 can be used without a significant intrinsic loss of efficiency for any area variation.

Chapter 7

Evaluation of the cell designs in a three-stack system

7.1 System design of the three-stack string system

Using the twenty-four cell designs from Section 5.3 on page 81, 40-cell stacks are modeled. Three stacks are connected in series electrically to boost the battery voltage. The electric series connection of stacks is called a string.

The three stacks are assumed to be placed vertically stacked in a rack. Common circuitry is carried out using a pipe diameter which is 50 % larger than the manifold diameter of the particular cell design. The pipe length from tank to rack is estimated to be 3 m. In terms of orifices, tank outlet (k_L=0.42), three 90°-bends (k_L=0.3, each) and sensors (k_L=2) are considered in the piping section from the tank to the first stack.

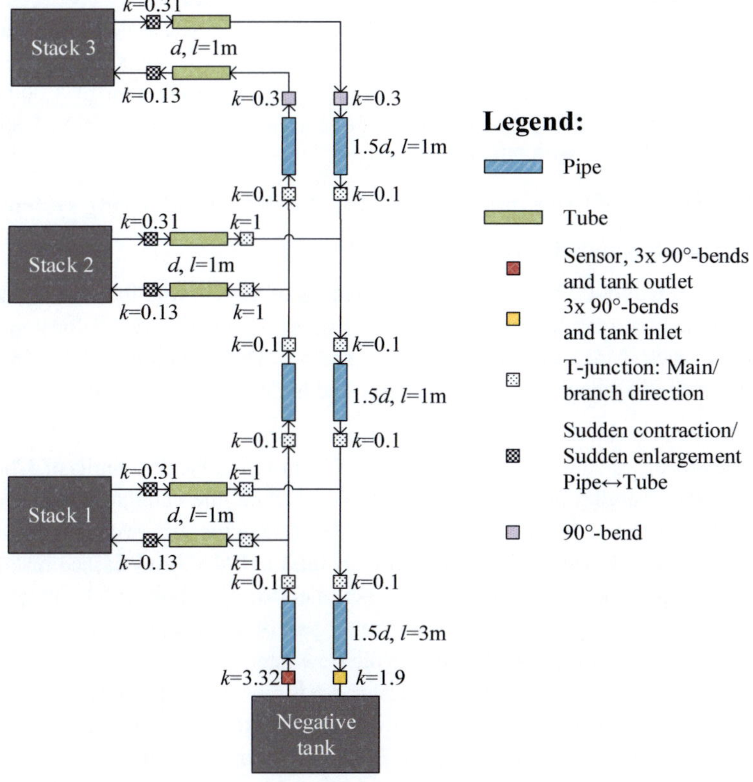

Figure 7-1: Piping plan of the generic three-stack VRFB system

The first and second stack is connected to the common circuitry via T-junctions. In the main flow direction, they introduce a loss coefficient of 0.2 which is distributed equally on input and output of the T-junction. In branch flow direction, an additional loss coefficient of one is considered. The last stack is connected to the common circuitry via a single 90°-bend with a loss coefficient of 0.3.

All stacks are connected to the piping using flexible tubes with a length of 1 m and a diameter which corresponds to the manifold diameter of the particular design. The sudden contraction from pipe to tube diameter causes a loss coefficient of 0.13, calculated with Eq. (2-92) on page 54, and is placed at the stack electrolyte inlet. The sudden expansion of tube to pipe causes a loss coefficient of 0.31, calculated with Eq. (2-91) on page 54, and is placed at the stack electrolyte outlet.

Figure 7-1 shows the complete hydraulic circuit. Compared to the single-stack study, the nominal pump capacity is scaled by a factor of three. The relative pump efficiency remains the same.

To obtain a comparable energy to power ratio, tank volume is also scaled by a factor of three. Further, it is adapted according to the electrode area, deployed by the design. Hence, the three stack-strings using the cells with an electrode area of 1,000 cm², 2,000 cm², 3,000 cm² and 4,000 cm² are assigned to tank volumes of 750 L, 1,500 L, 2,250 L and 3,000 L per half-side, respectively.

7.2 Comparison of two sample designs for the three-stack string system

7.2.1 Dynamic simulation results

For illustration purposes, dynamic simulation results of a cycle with the lowest studied and the nominal current density are shown in Figure 7-2 and Figure 7-3 for a system with a three-stack string using cell designs 1.1 and 4.6. The applied flow factor in each case is selected according to the process, described in Section 6.3.1 on page 86.

7.2.1.1 Operation with the lowest studied current density

If the battery is operated with a current density of 25 mAcm⁻², the SoC limits of 5 % and 90 % limit the charging and discharging process. Hence, the cells do not reach the voltage limits, as shown in Figure 7-2 a) and b). Referred to the string voltage, the lower voltage limit is 132 V, whereas the upper voltage limit is 198 V (gray dashed lines).

Naturally, pump capacity is not completely exploited for operation with a low current density, as shown in Figure 7-2 c). For the better part of the operation time, the lower flow rate limitation of the pump determines the flow rate.

Although the channel geometry factor of design 4.6 is more than three times larger than the channel geometry of design 1.1, the equivalent shunt currents of both designs hardly differ from each other, as shown in Figure 7-2 d). This can be explained by the impact of the external piping network. For design 1.1, the tube diameter is 30 mm, while the pipe diameter is 45 mm. For design 4.6, the tube diameter is 60 mm, while the pipe diameter is 90 mm. Hence, for design 1.1, the external hydraulic network represents a larger ionic resistance towards the shunt currents. This phenomenon is further analyzed in Section 7.3.5 on page 112.

In the worst-case scenario for cell design 1.1, an external current of only 25 A is applied. In this case, the shunt currents waste up to 15 % of this current, as shown in Figure 7-2 e). For design 4.6, the externally applied current is four times larger, due to the fourfold larger electrode area. Thus, the maximum share of lost current is only 4 %.

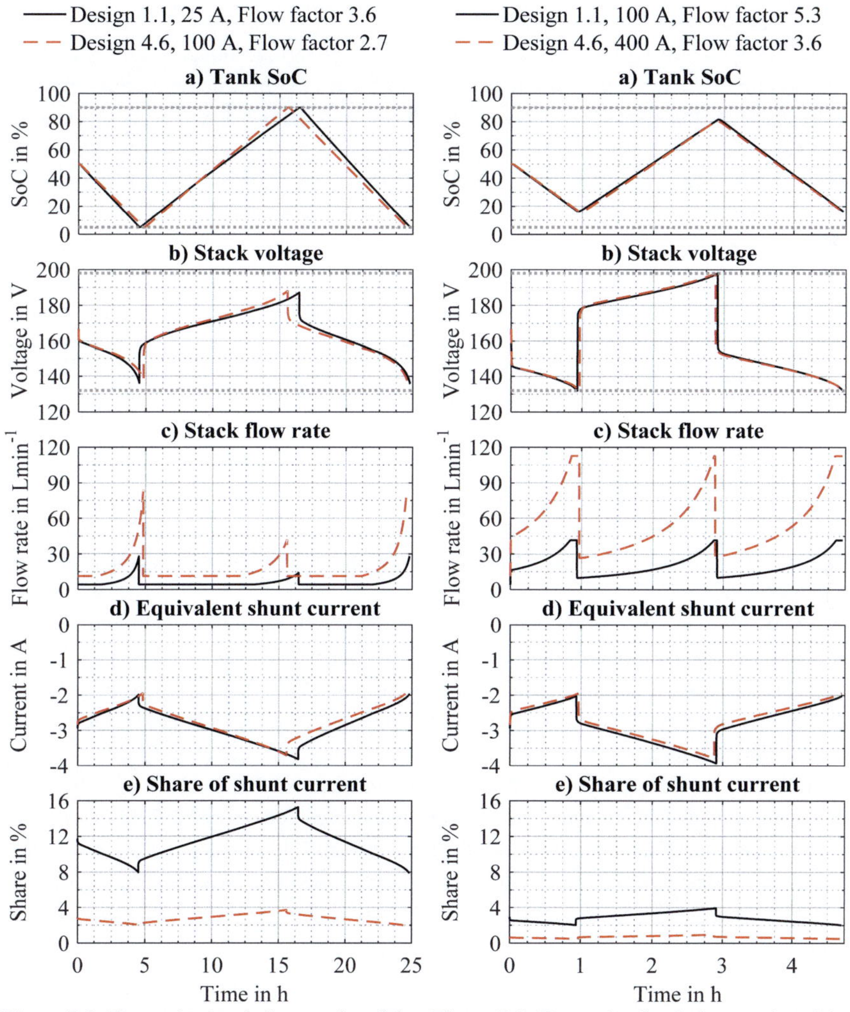

Figure 7-2: Dynamic simulation results of the three-stack system for a current density of 25 mAcm^{-2}

Figure 7-3: Dynamic simulation results of the three-stack system for the nominal current density of 100 mAcm^{-2}

Shunt currents cause a significant temporal deviation between the cycles simulated with the two designs. Due to the massive current loss during the charging process, it takes significantly longer to charge the battery with cell design 1.1. During the discharging process, shunt currents increase the effective discharging current, which accelerates the discharging of the battery. Naturally, the combination of the two effects substantially lowers the efficiency. Consequentially, operating a multi-stack string with a low current density should be avoided, in particular with a small electrode area.

7.2.1.2 Operation with nominal current density

During the operation with nominal current density, the string voltage limits bound the charging and discharging operation, as shown in Figure 7-3 b). The target SoC of 80 % during the charging process is reached, as shown in Figure 7-3 a). For both designs, pump capacity is fully utilized, as shown in Figure 7-3 c). The trend of the equivalent shunt current hardly deviates from the one of the lowest studied current density, as shown in Figure 7-3 d). However, now that a larger current density is applied, the ratio between externally applied current and equivalent shunt current improves, as shown in Figure 7-3 e). While design 1.1 wastes up to 3.9 % of the externally applied current, this share is only 0.9 % for design 4.6. Hence, for operation with nominal current density, shunt currents play a minor role as a loss mechanism in a three-stack string, in particular for a large electrode area.

7.2.2 Loss distribution

In this section, the accumulated losses during the four previously simulated cycles are analyzed. The values are referred to the discharge capacity to allow for a comparison of the two designs.

For the operation at low current densities, Coulomb losses, namely vanadium crossover and shunt currents, dominate the total losses, as shown in Figure 7-4. For design 1.1, shunt current losses dominate over crossover losses. With design 4.6, shunt current losses are reduced by 78 %. Compared to crossover losses, they only contribute half as strong to the total losses.

For the operation at nominal current density, ohmic losses contribute the biggest share to the total losses. Due to the smaller flow factor used for design 4.6, losses due to concentration overpotential are higher than for design 1.1. Despite the 32 % smaller flow factor for design 4.6, it has a higher pump energy demand, as shown in Figure 7-4. The higher pump energy demand is caused by the larger hydraulic resistance of this design.

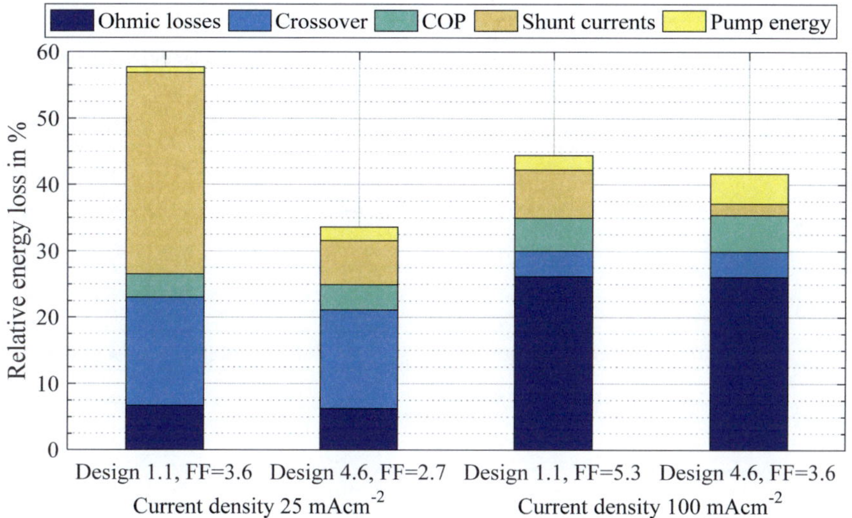

Figure 7-4: Relative energy loss of designs 1.1 and 4.6 in a system using a three-stack string for a cycle with 25 mAcm^{-2} and 100 mAcm^{-2} (FF = Flow factor)

7.2.3 Efficiencies of the sample designs

The Coulomb efficiency is the only efficiency which changes noteworthy for a three-stack string compared to a system using a single-stack, as shown in Figure 7-5.

Due to its larger electrode area and thus its larger current carrying capability, the Coulomb efficiency of design 4.6 is significantly higher than the Coulomb efficiency of design 1.1. Naturally, this affects energy and system efficiency accordingly. Shunt currents significantly lower the efficiency of design 1.1 compared to design 4.6.

The dependency of all efficiencies on the flow factor is the same as extensively laid out in Section 6.4.2 on page 91.

7.2.4 Discharge capacities of the sample designs

Again, the discharge capacity shows a strong dependence on the applied flow factor and the applied current density. However, the general correlations are the same as laid out in Section 6.4.3 on page 93.

For design 1.1, the discharge capacity for the lowest current density is strongly affected by shunt currents, as shown in Figure 7-6 a). Hence, for a reasonable flow factor, it is lower than the discharge capacity of the next larger studied current density, which is counter-intuitive. Besides the negative effect of shunt currents on the operation with a low current density, the discharge capacity is not affected by the electrical series connection of three stacks.

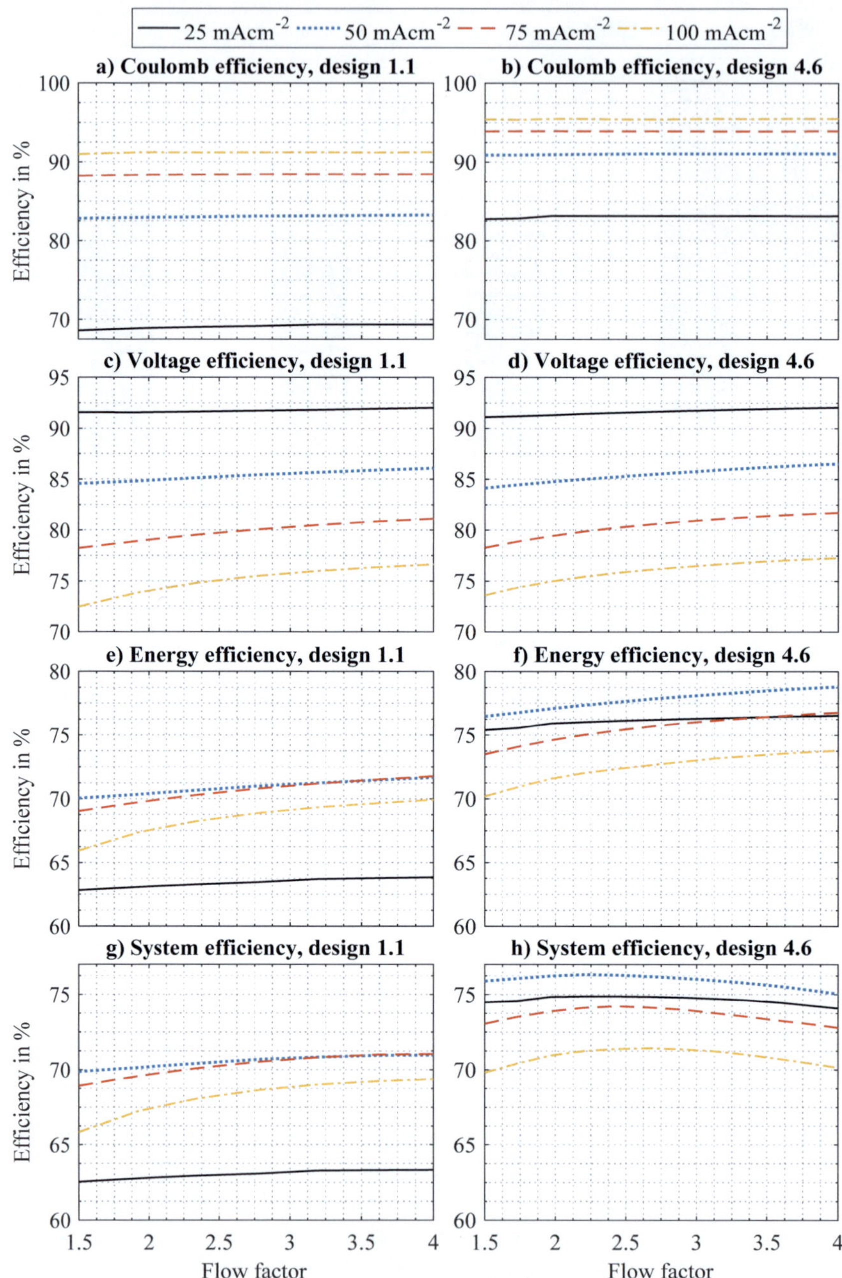

Figure 7-5: Coulomb, voltage, energy and system efficiency of design 1.1 and design 4.6 in a three-stack string over the flow factor for different current densities.

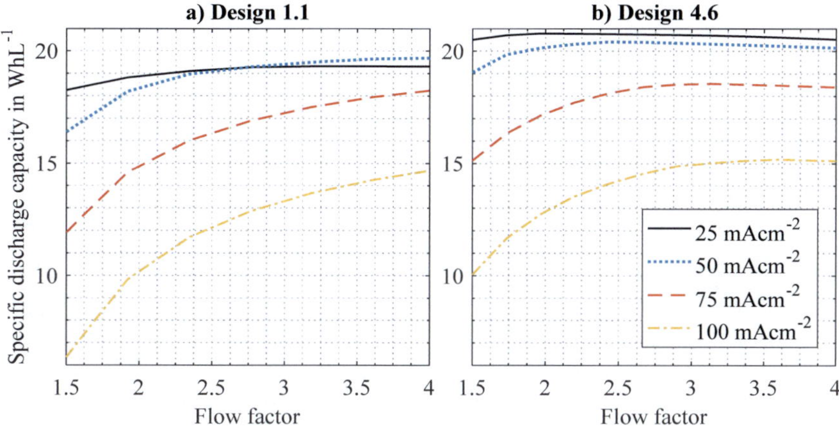

Figure 7-6: Specific discharge capacity over flow factor and current density for design 1.1 and design 4.6 in a three-stack string

7.3 Evaluation of the twenty-four cell designs in a three-stack string

7.3.1 Coulomb efficiency

The impact of channel geometry factor and electrode area on Coulomb efficiency is stronger for the three-stack string than for the single stack, as shown in Figure 6-6 on page 95 and in Figure 7-7. Both, a larger channel geometry factor and a larger electrode area significantly increase the Coulomb efficiency. Unfortunately, both mechanisms face a saturation effect. Hence, it is hardly possible to boost the Coulomb efficiency of a 1,000-cm^2 electrode to the level of a 4,000-cm^2 electrode only by means of an increased channel geometry factor.

7.3.2 Voltage efficiency

For the sake of completeness, the voltage efficiency of the three-stack string for a cycle with the nominal current density is studied as well. However, if we compare the derived results of the three-stack string, shown in Figure 7-8, with the corresponding results of the single-stack system, shown in Figure 6-7 on page 95, we see that the electric series connection does not affect the voltage efficiency.

7.3.3 Optimal flow factor

Compared to the hydraulic resistance of the stack, the hydraulic resistance of the external hydraulic circuit is significantly smaller. Further, we just learned that the voltage efficiency is not affected by serially connecting several stacks electrically. Compared to the single-stack system, the first fact implies a similar pump power demand of the three-stack string. The second fact implies a similar concentration overpotential.

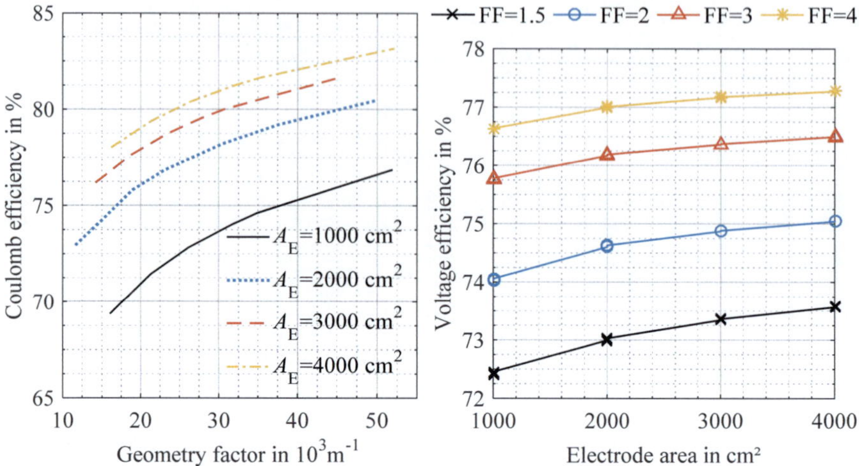

Figure 7-7: Coulomb efficiency for the three-stack string at 25 mAcm^{-2} over channel geometry factor and electrode area

Figure 7-8: Voltage efficiency for the three-stack string at 100 mAcm^{-2} over electrode area and flow factor

Table 7-1: Deployed flow factors for all twenty-four designs in a three-stack string for lowest and nominal current density. In brackets: Optimal flow factor in terms of RTSE, if applied flow factor deviates from optimal flow factor

Area variation	Electrode area	Channel design variation					
		1 short wide	2 short medium	3 short narrow	4 long wide	5 long medium	6 long narrow
		Current density 25 mAcm^{-2}					
1	1000 cm²	4.0	4.0	3.6	4.0	4.0	3.6
2	2000 cm²	3.4	3.1	3.1	3.4	3.1	2.8
3	3000 cm²	3.1	2.8	2.8	2.8	2.8	2.6
4	4000 cm²	2.9	2.7	2.4	2.7	2.4	2.4
		Current density 100 mAcm^{-2}					
1	1000 cm²	5.3 (4.0)	5.3 (4.0)	5.3 (4.0)	5.3 (4.5)	5.3 (4.0)	5.3 (4.0)
2	2000 cm²	4.4 (3.8)	4.4 (3.4)	4.4 (3.4)	4.4 (3.8)	4.4 (3.4)	4.4 (3.1)
3	3000 cm²	3.9 (3.1)	3.9 (3.1)	3.9 (3.1)	3.9 (3.1)	3.9 (3.1)	3.9 (3.1)
4	4000 cm²	3.6 (2.9)	3.6 (2.9)	3.6 (2.7)	3.6 (2.9)	3.6 (2.7)	3.6 (2.7)

As a consequence, the optimal flow factor, which is mainly a compromise between pump power and concentration overpotential, only varies little between the single-stack system and the system with a three-stack string, as shown in Table 7-1.

Again, for nominal current density, the optimal flow factor in terms of efficiency is not large enough to allow for a charging operation up to a tank SoC of 80 %. Hence, it is replaced by the flow factor which permits the desired operation. For these cases, the most efficient flow factors are given in brackets. In general, the optimal flow factor is again larger for a smaller electrode area and only little affected by the channel geometry.

7.3.4 Auxiliary losses

For the system using the three-stack string, similar hydraulic resistances and similar flow factors imply a similar pump energy demand, referred to the number of stacks. Hence, the values given in Table 7-2 for the three-stack string only deviate little from the values given in Table 6-4 on page 99 for the single-stack system. As observed for the single-stack system, a larger electrode and a longer and narrower channel lower the auxiliary efficiency. The former is because of the increased electrolyte demand, the latter is because of the increased channel hydraulic resistance. Averaged over all designs, the auxiliary efficiency of a three-stack system is 0.2 %-points lower than the one of the single-stack system.

Table 7-2: Auxiliary efficiency due to the pump power demand of all twenty-four designs in a three-stack string for a cycle with the nominal current density and the flow factors according to Table 7-1

		Channel design variation					
Area variation	Electrode area	1 short wide	2 short medium	3 short narrow	4 long wide	5 long medium	6 long narrow
1	1000 cm²	98.1 %	97.9 %	97.6 %	97.9 %	97.7 %	97.0 %
2	2000 cm²	97.9 %	97.6 %	97.3 %	97.8 %	97.2 %	96.8 %
3	3000 cm²	97.7 %	97.6 %	97.2 %	97.5 %	97.3 %	96.7 %
4	4000 cm²	97.4 %	97.3 %	96.8 %	97.2 %	96.9 %	96.1 %

7.3.5 Shunt currents

In a string consisting of several stacks, cell geometry parameters no longer exclusively determine the shunt currents. In fact, the geometry of the external hydraulic circuit, namely diameters and lengths of pipes and tubes, has a significant impact.

In Figure 7-9, the equivalent shunt current under no load conditions for a tank SoC of 50 % are grouped according to the tube and pipe diameters of the respective designs. Each group shows the well-known hyperbola regarding the dependence on the channel geometry factor. In this case, the smaller combinations of tube and pipe diameter correspond to a smaller electrode area. One could also choose smaller tube and pipe diameters for the larger electrodes, but this will negatively affect the homogenous distribution of the pump flow rate on the individual stacks, as the author extensively shows in reference [9].

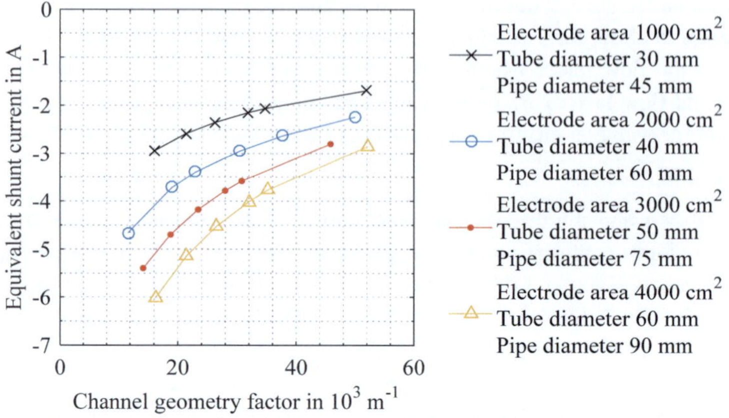

Figure 7-9: Equivalent shunt currents of all designs in a three-stack string
for a tank SoC of 50 % under no load conditions.

Figure 7-9 demonstrates that the equivalent shunt current can easily reach absolute values of 6 A and beyond, if several stacks are connected in series electrically.

Besides negatively affecting the efficiency, shunt currents cause additional problems. The graphite felt used as electrode has a typical conductivity in the range of 300 Sm^{-1} [90]. For the bipolar plate, the conductivity is around 5,000 Sm^{-1} [91]. Hence, the electric conductivity of electrode and bipolar plate is approximately ten times larger than the conductivity of the electrolyte, which is 23.7 Sm^{-1} for the negative electrolyte and 37.1 Sm^{-1} for the positive electrode, both in a SoC of 50 %. This can be calculated from the Eqs. (2-32) and (2-33) on page 28.

Consequentially, shunt currents start and end at the very first point, where graphite felt or bipolar plate touches the electrolyte in the particular cell. Hence, the local current density in this spot is substantially higher than expected in regular operation. Due to the ratios of the conductivities, we have to expect that all shunt currents originate from a few square centimeters or even square millimeters of area. This very high current density might accelerate the ageing of the concerned area. In [37], the authors identify high overpotentials in the relevant areas which trigger side reactions like oxygen evolution and carbon corrosion. This effect is another reason why we have to carefully address the shunt current phenomenon while designing a VRFB cell.

7.3.6   Final evaluation using the round-trip system efficiency

7.3.6.1 Lowest studied current density

For a current density of 25 mAcm^{-2}, the efficiencies of the twenty-four designs vary by up to 11.6 %-points, as shown in Table 7-3. A larger electrode clearly yields a higher efficiency. In addition, a longer and narrower channel positively affects the efficiency. Hence, design 4.6 is the most efficient design for a low current density.

7.3.6.2 Nominal current density

For nominal current density, the efficiencies of the different designs vary significantly less. The difference between most and least efficient design is only 1.7 %-points. The general trend of larger cells and narrower, longer channels to yield a higher efficiency is no longer thoroughly valid in this case. The reason for this is that the high applied current density increases the required flow rate. This is disadvantageous for a cell with a large electrode, because the hydraulic resistance of its channels causes a larger pump power demand in this case.

Three almost equally efficient designs can be identified for the nominal current density, of which design 3.5 is the most efficient one. However, its advantage is only 0.05 %-points.

Table 7-3: RTSE for two different current densities for all twenty-four designs in a three-stack string. Bold: Highest RTSE for the particular current density.

Area variation	Electrode area	Channel design variation					
		1 short wide	2 short medium	3 short narrow	4 long wide	5 long medium	6 long narrow
Current density 25 mAcm^{-2}							
1	1000 cm²	63.3 %	65.2 %	67.5 %	66.5 %	68.0 %	70.0 %
2	2000 cm²	66.4 %	69.8 %	71.0 %	69.0 %	71.9 %	73.0 %
3	3000 cm²	69.1 %	70.4 %	72.1 %	71.4 %	72.5 %	73.9 %
4	4000 cm²	70.6 %	71.8 %	73.3 %	72.7 %	73.7 %	**74.9 %**
Current density 100 mAcm^{-2}							
1	1000 cm²	69.2 %	69.6 %	69.9 %	69.9 %	70.2 %	70.2 %
2	2000 cm²	69.9 %	70.6 %	70.6 %	70.4 %	70.8 %	70.7 %
3	3000 cm²	70.4 %	70.6 %	70.7 %	70.7 %	70.9 %	**70.9 %**
4	4000 cm²	70.5 %	70.7 %	70.7 %	70.8 %	70.9 %	70.7 %

7.3.6.3 Current-weighted average RTSE

Table 7-4: Current-weighted average RTSE values for all twenty-four designs

Area variation	Electrode area	Channel variation					
		1 short wide	2 short medium	3 short narrow	4 long wide	5 long medium	6 long narrow
1	1000 cm²	69.5 %	70.3 %	71.1 %	70.8 %	71.3 %	71.9 %
2	2000 cm²	70.7 %	72.0 %	72.3 %	71.7 %	72.7 %	72.9 %
3	3000 cm²	71.6 %	72.1 %	72.6 %	72.4 %	72.8 %	73.2 %
4	4000 cm²	72.1 %	72.5 %	72.9 %	72.8 %	73.1 %	**73.3 %**

The efficiencies of the individual current densities are converted into an aggregated efficiency value by means of current-weighted averaging. With a small advantage of 0.1 %-points, cell design 4.6 turns out to be the most efficient design. This is not surprising, as it combines the largest electrode area with the largest channel geometry factor. While its channel geometry leads to a minor efficiency loss when operating with nominal current density, it is of big advantage when operating with lower current densities. In addition, design 4.6 yields the lowest shunt currents and thus minimizes corrosion caused by this phenomenon.

7.4 Conclusion

Serially connecting several flow battery stacks electrically introduces two major disadvantages. First, a multi-stack string will always be less efficient than a single-stack. There is no internal effect enabling an efficiency gain, while shunt currents compellingly increase in such an arrangement.

If one compares the most efficient single-stack system to the most efficient three-stack string system, the efficiency loss ranges from 1.6 to 3.3 %-points, as shown in Figure 7-10. This is despite the selection of twice the electrode area and a 39 % increase in channel geometry factor for the cell design used in the three-stack string.

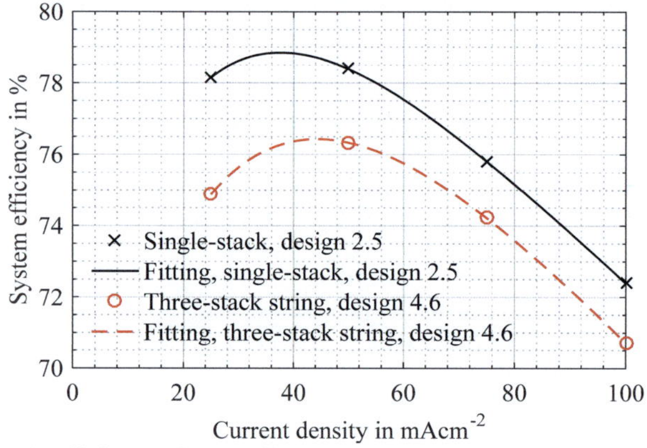

Figure 7-10: Efficiency of most efficient single-stack and three-stack string system

7.5 Implications for grid connection

Regardless of the efficiency loss, the higher DC voltage of a multi-stack string facilitates the grid connection of the battery. This might also enable a higher efficiency of the PCS. We can estimate the required PCS efficiency for a three-stack string to operate as efficient as a single-stack as follows.

Two systems are compared to each other. The first system consists of six stacks, each comprising 40 cells using cell design 2.5 (highest efficiency in a single-stack system, electrode area 2,000 cm^2) in electric parallel connection. The second system consists of three stacks, each comprising 40 cells using cell design 4.6 (highest efficiency in a three-stack string, electrode area 4,000 cm^2) in electric series connection. Both systems have a comparable total electrode area and thus a comparable power rating. For the sake of simplicity, it is assumed that the system with six parallel stacks is equally efficient as the single-stack system.

The system with six parallel stacks has to be connected to the power grid using a PCS with a low input DC voltage. Table 7-5 gives the efficiency of a typical PCS, η_{PCS}, for the voltage level of a 40-cell stack [92]. The product of the system efficiency of design 2.5 with the PCS efficiency, η_{PCS}, gives the total efficiency of the single-stack system, η_{Total}.

If we divide this efficiency by the system efficiency of design 4.6 in a three-stack string, we yield the efficiency of the three-stack PCS which is required to operate the three-stack string as efficiently as the single stack.

Table 7-5: Computation of required PCS round-trip efficiency to compensate for the loss in Coulomb efficiency due to the electric series connection of three stacks.

Current density in mAcm^{-2}	25	50	75	100
η_{Sys}, design 2.5, six stacks in parallel	78.2 %	78.4 %	75.8 %	72.4 %
η_{PCS}, E_{DC} = 54 V	95.0 %	94.0 %	92.5 %	91.5 %
η_{Total}, design 2.5, six stacks in parallel	74.3 %	73.7 %	70.1 %	66.2 %
η_{Sys}, design 4.6, three-stack string	74.9 %	76.3 %	74.2 %	70.7 %
Required η_{PCS}, E_{DC} = 162 V	**99.2 %**	**96.6 %**	**94.5 %**	**93.6 %**

It is obvious that for an operation with 25 mAcm^{-2}, the three-stack string can hardly obtain the same total efficiency than the single stack. This is because a PCS round-trip efficiency of 99.2 % is rather utopic at this input voltage level. When operating with nominal current density, the PCS for the three-stack string needs to have an additional 2.1 %-points of efficiency compared to the single stack. Considering the three-fold increased input DC voltage, this appears to be possible. Hence, if operated with a sufficiently high current density, it appears to be possible to operate a three-stack string as efficiently as a single stack.

However, if a sufficiently high PCS efficiency is available for the low DC voltage of a single VRFB stack, it would no longer be possible to obtain the required efficiency gains due to the higher DC voltage. In addition, resigning from serially connecting several stacks electrically is beneficial for the system lifetime due to reduced corrosion by shunt currents.

It thus can be concluded that developing a high efficient PCS with a low input DC voltage is of high interest for the system design of redox flow batteries. It is noteworthy that several companies and research institutes are currently active in the field of high efficient low voltage AC/DC converters [93].

Chapter 8

Flow rate optimization

8.1 Current state of science

The simplest flow rate control strategy (FRCS) of a VRFB is deploying a constant flow rate for all SoCs and currents. Another simple approach is to adapt the flow rate according to the instantaneously derived stoichiometric requirements of the electro-chemical reactions. These requirements are given by Faraday's first law of electrolysis. To compensate for imperfections and losses, the actual flow rate has to be larger than the stoichiometric one. Hence, the instantaneous stoichiometric flow rate is multiplied by a constant factor, commonly referred to as the 'flow factor' [30].

The two approaches mentioned above have already been studied by NASA for the iron-chromium flow battery in 1982 [94]. More than 30 years later, a similar comparison is presented for the VRFB [30]. In terms of discharge capacity and system efficiency, a variable flow rate using the scaled instantaneous stoichiometric flow rate is found to be advisable for the VRFB. One drawback of the presented study is the over-estimation of the pump efficiency. In [30], a constant pump efficiency of 80 % is assumed. This is more than twice as high as the best efficiency point of real pumps, as shown in Section 2.12.6 on page 54.

In [17, 95], a third FRCS is presented, representing a simple and straightforward optimization approach. If we charge the battery with a constant power, we have to adapt the flow rate depending on the SoC with the goal of maximizing the applied charging current. This is reasonable because the charging progress depends on the applied current, not on the applied power. Hence, the claim for a maximum current at a given power implies a low charging voltage, which benefits the efficiency. If we discharge the battery with a constant power, we adapt the flow rate depending on the SoC to minimize the absolute value of the discharging current.

If we charge the battery with a constant current, the goal is to minimize the input power. If we discharge the battery with a constant current, the goal is to maximize the absolute value of the output power. This methodology is also used in [96]. Herein, the impact of temperature and pipe diameter on the efficiency is studied in addition.

However, due to additionally considered loss mechanisms, namely shunt currents and vanadium crossover, the previously presented approach of minimizing and maximizing current and power, respectively, is no longer adequate.

In addition to the presented model-based studies, some experimental studies have been published as well. By increasing the flow rate at pre-defined cell or stack voltage levels, a good compromise between system efficiency and discharge capacity can be yielded [97]. Herein, an experimental study is carried out with a kW-class VRFB. However, the values of the pre-set voltage limits and the values of the applied low and high flow rates are not varied. Hence, it can be assumed that this method represents a simple approach for finding an acceptable compromise between system efficiency and discharge capacity, but significant optimization potential remains unused.

Another approach to reduce the average pump power demand is to apply a pulsating electrolyte flow [98]. Using a 20-cm^2 lab-scale flow cell, it is shown that significant energy savings are achievable with different on- and off-periods of the pumps. However, a pulsating flow rate exposes the stack to transient mechanical loads, which might reduce the lifetime and might increase the risk of leakages. Further, the method comprises a large number of variables, namely pump on- and off-times and applied flow rates for both, the charging and discharging process. Most likely, the optimal parameter set will additionally vary with SoC and applied charging and discharging current. Hence, an extensive optimization process is required to bring this method into practice.

8.2 Minimally required flow rate – Faraday's law

The applied flow rate is the most important operational parameter in a flow battery. Faraday's first law of electrolysis represents the lower absolute limit of the flow rate in steady state. Applied to the VRFB, Faraday's law states that the available number of vanadium ions, partaking in the specific electrochemical reaction, has to be equal to the number of electrons involved in the applied electric current. For the negative and positive half-cell, this results in the Eqs. (8-1) and (8-2).

$$Q_{C-} = \begin{cases} \dfrac{I_C}{Fc_{3T-}}, & \text{for charging} \\ \dfrac{|I_C|}{Fc_{2T-}}, & \text{for discharging} \end{cases} \tag{8-1}$$

$$Q_{C+} = \begin{cases} \dfrac{I_C}{Fc_{4T+}}, & \text{for charging} \\ \dfrac{|I_C|}{Fc_{5T+}}, & \text{for discharging} \end{cases} \tag{8-2}$$

For a well-balanced electrolyte, we can express the vanadium species by tank SoC and total vanadium concentration, as shown in Eq. (8-3).

$$Q_C = \begin{cases} \dfrac{I_C}{F(1 - SoC_T)c_V}, & \text{for charging} \\ \dfrac{|I_C|}{FSoC_Tc_V}, & \text{for discharging} \end{cases} \tag{8-3}$$

Obviously, the minimally required flow rate depends on tank SoC and applied current. It is proportional to the magnitude of the applied electric current. Regarding the SoC, we have to take the operation mode into account. During the charging process, a high SoC requires a significantly larger flow rate than a low SoC, as shown in Figure 8-1. This can be easily explained by evaluating the composition of the two electrolytes, which are fed to the cells. A high SoC corresponds to a high concentration of V^{2+} ions in the negative electrolyte and a high concentration of VO_2^+ ions in the positive electrolyte. Consequentially, the concentrations of V^{3+} and VO^{2+} ions in the respective electrolyte are rather low. If we want to charge the electrolytes to an even higher SoC, a sufficient number of V^{3+} and VO^{2+} ions still has to be present in the cells. The number of available ions is the product of ionic concentration and volumetric flow rate. As the concentration decreases further, the volumetric flow rate has to rise.

At a low SoC, the situation is contrarily. In this case, the charging process requires a small flow rate, while the discharging process requires a high flow rate.

The flow rate computed with Faraday's law is only applicable, if the Coulomb efficiency is unity, i.e. not a single electron is lost due to any parasitic process. In reality as well as in the model, this does not hold true. Shunt currents and vanadium crossover need to be considered. Therefore, the applied flow rate has to be higher than the stoichiometric one.

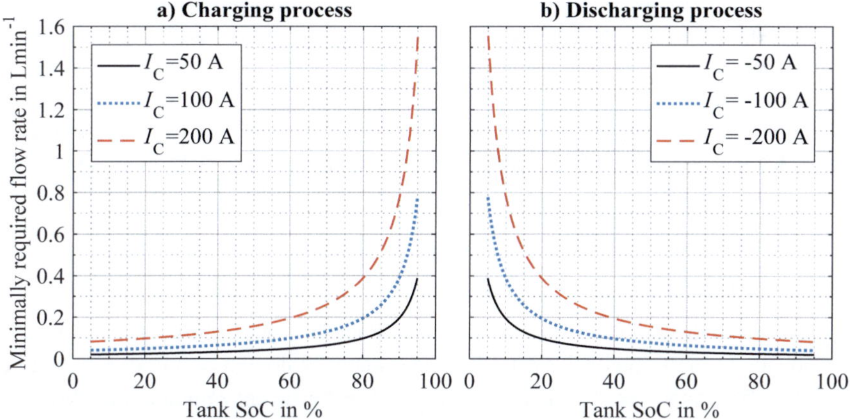

Figure 8-1: Minimally required flow rates over tank SoC according to Faraday's first law of electrolysis for charging and discharging process with different currents (cv=1,600 molm^{-3})

8.3 Flow rate optimization basics

The applied flow rate always affects system efficiency and discharge capacity simultaneously. For the system efficiency, we mainly have to consider the antagonists pump energy demand and energy loss due to concentration overpotential. The flow rate affects the discharge capacity if the cell or stack voltage limits the battery operation, because the flow rate has a strong impact on the cell voltage, as shown in the Eqs. (5-1) and (5-2) on page 80.

A higher flow rate lowers the battery voltage during the charging process and thus allows for a higher tank SoC to be reached until the upper voltage limit is violated. During the discharging process, a higher flow rate boosts the discharging voltage and increases the time until the lower voltage limit is violated. Hence, the discharging process finishes at a lower tank SoC. Both, a higher tank SoC at the end of the charging process and a lower tank SoC at the end of the discharging process, increase the discharge capacity. Hence, the higher the flow rate, the larger the useable SoC range. However, in practice, this correlation is hindered by two reasons. First, the pump capacity is limited. Secondly, the pump power demand increases quickly with an increasing flow rate, as illustrated in Figure 2-23 on page 56. Hence, system efficiency is negatively affected. For very large flow rates, it is most likely that the additional pump energy demand completely consumes the gained additional discharge capacity.

In this work, the maximization of system efficiency and discharge capacity is considered as two separate optimization objectives. It is of course possible to combine both objectives. However, without knowing the intended battery use-case, it is not possible to decide whether a certain loss in system efficiency is economically compensated by the additionally provided discharge capacity or vice versa. Hence, the two extreme optimization objectives are studied to give the boundaries for any combination of the two.

Flow rate optimization is carried out for cell design 2.5, which yields the highest average RTSE of all studied designs in a single-stack system. Hence, electrode area is 2000 cm², channel length is 1,129 mm and channel width is 10 mm. A tank volume of 1,000 L per half-side is deployed. Again, the method of deriving the RTSE by means of simulating a successive charging and discharging process is applied. A pre-discharging process is carried out before the actual cycle.

8.4 Constant flow rate

8.4.1 Methodology

In terms of system complexity and control, a constant flow rate is the simplest approach. One advantage of this method is that we can design the hydraulic circuit for one specific flow rate. We can select the pumps to have their best efficiency point at this flow rate and no inverter is required to run the pump motor. Furthermore, we do not require flow rate or pressure sensors, which lowers the system costs.

While determining the constant flow rate, Faraday's law has to be obeyed as well. The constant flow rate has to be large enough to enable both the charging and discharging operation towards the highest and lowest SoC, respectively, to be reached with the nominal current. If the SoC limits are non-symmetrical to a SoC of 50 %, we have to select the larger of the two flow rates.

A VRFB using cell design 2.5 has a nominal current of 200 A. It shall be possible to obtain a tank SoC of 80 % during a charging process with this current. According to Eq. (8-3), the minimally required flow rate for the deployed electrolyte with a total vanadium concentration of 1.6 molL^{-1} for reaching a SoC of 80 % with a charging current of 200 A is 0.389 Lmin^{-1}. This flow rate is denoted as the basic flow rate. Nevertheless, it is possible to deviate from this flow rate in order to increase system efficiency or discharge capacity. In this work, the deviating flow rates are referred to the basic flow rate by using the scaling or flow factor, *FF*, as shown in Eq. (8-4). This simplifies the study of reasonable flow rate ranges.

$$Q_C = FF \cdot 0.389 \text{ Lmin}^{-1} \tag{8-4}$$

In contrast to the conventional variable FRCS, which is presented in the next section, the flow factor for the constant FRCS can be set to values below one. However, any flow factor below one will result in failing to obtain a tank SoC of 80 % during a charging operation with the nominal current.

The upper boundary of the studied flow factors can be limited by some simple considerations. For a 40-cell stack, the basic flow rate is 15.6 Lmin^{-1}. For design 2.5, the nominal pump capacity is 67.8 Lmin^{-1}. Hence, the largest reasonable flow factor is 4.35. Consequently, simulations with a flow factor between 0.4 and 4.35 are conducted with different current densities between 20 and 100 mAcm^{-2} in steps of 10 mAcm^{-2}.

The advantage of using a pump optimized for the particular flow rate is explicitly considered. Hence, the best efficiency point of the flow-rate-dependent efficiency $(max(\eta_{Pump}) = 34.6 \text{ %})$, as shown in Figure 2-21 on page 55, is used as the constant pump efficiency.

8.4.2 Simulation results

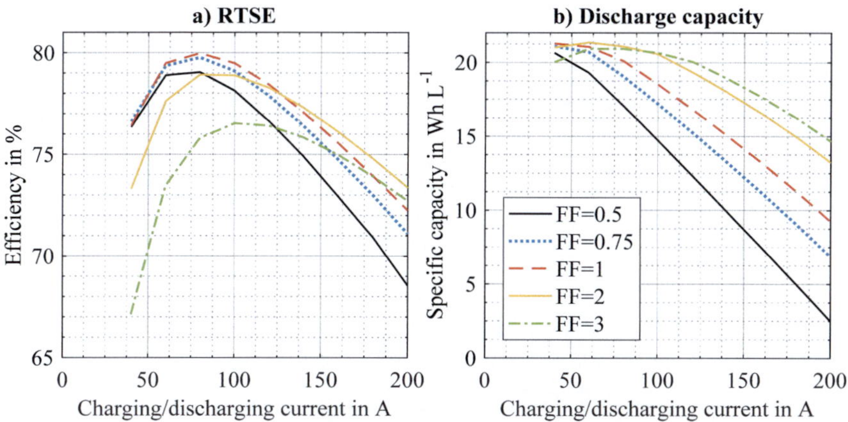

Figure 8-2: RTSE and specific discharge capacity over applied current for design 2.5 and the constant FRCS with different flow factors (FF)

For a smaller current, a smaller flow factor yields a higher efficiency, as shown in Figure 8-2 a). A larger flow factor dramatically worsens the efficiency. For a large current, the situation is contrary. In this case, a smaller flow factor yields a lower efficiency.

In terms of capacity, a larger flow factor yields a larger discharge capacity, as shown in Figure 8-2 b). This is valid as long as the system efficiency does not decrease too strongly due to the increased pump power demand. For smaller currents, the superimposed SoC limits additionally limit the capacity gain of a larger flow factor. The charging process is ended when the tank SoC reaches a value of 90 %, even if the upper stack voltage limit is not yet violated. The same applies to the discharging current in an analogue manner, for a lower tank SoC limit of 5 %.

A further increase of the flow factor only results in a capacity gain, if it increases the efficiency. For a current of 40 A, increasing the flow factor from two to three drastically lowers the RTSE. Hence, despite the general correlation, it is also possible for a larger flow factor to yield a lower discharge capacity.

We can conclude from Figure 8-2 that different flow factors are optimal for different currents. Unfortunately, the constant FRCS does not permit different flow factors. Hence, the current-weighted average of the RTSE, ØRTSE, is calculated as shown in Eq. (8-5) to identify the optimal flow factor for the constant FRCS. The values of ØRTSE are calculated individually for all the different flow factors, as shown in Figure 8-3 a).

$$\emptyset RTSE = \frac{\sum_{i=1}^{N_j} j_i \eta(j_i)}{\sum_{i=1}^{N_j} j_i} \tag{8-5}$$

Wherein:

$\eta(j_i)$ RTSE for the considered current density j_i (%)
j_i i^{th} considered current density   (Am^{-2})
N_j Number of different considered current densities   (-)

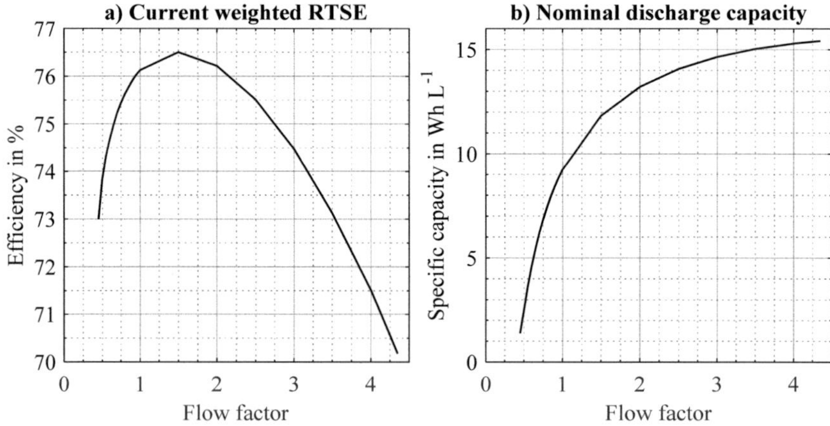

Figure 8-3: Current-weighted average RTSE and nominal discharge capacity over the flow factor for the constant FRCS of design 2.5.

The highest current-weighted RTSE identifies the constant flow factor which is most suitable over the whole operational area. In terms of capacity, the nominal discharge capacity yielded by a specific flow factor for the nominal current is evaluated, as shown in Figure 8-3 b).

With a $\varnothing RTSE$ of 76.5 %, a flow factor of 1.5 yields the highest current-weighted average RTSE. However, this flow factor yields a nominal discharge capacity of only 11.8 WhL^{-1}.

The nominal capacity increases with the flow factor, as long as the pump capacity is not maxed out, as shown in Figure 8-3 b). Hence, the maximum reasonable flow factor of 4.35 is deployed to maximize the nominal discharge capacity. This flow factor yields a nominal discharge capacity of 15.4 WhL^{-1}, while yielding a current-weighted average RTSE of 70.2 %. Hence, the additional nominal discharge capacity of 3.6 WhL^{-1} (+31 %) reduces the current-weighted average RTSE by 6.3 %-points.

8.5 Variable flow rate – Conventional approach

8.5.1 Methodology

The principal advantage of a variable flow rate is its ability to consider the specific requirements of different operational conditions of the battery, e.g., different SoCs and currents. The conventional variable FRCS constantly computes the stoichiometrically required flow rate for the present combination of tank SoC and current, as shown in Eq. (8-6) [94]. Due to Coulomb losses and for optimization purposes, this value is scaled with a flow factor larger than one. In this work, the flow factors for the charging and discharging process are always identical. The conventional variable FRCS is comprehensively presented in [17, 30].

$$Q_C = \begin{cases} \dfrac{FF \cdot I_C}{F(1 - SoC_T)c_V}, & \text{for charging} \\[2em] \dfrac{FF \cdot I_C}{F SoC_T c_V}, & \text{for discharging} \end{cases} \tag{8-6}$$

Similar to the constant FRCS, we can either optimize the flow factor to maximize system efficiency or discharge capacity. In this work, flow factors between 1.5 and 8 are studied in steps of 0.25 for the conventional variable FRCS.

Here, a flow factor below one is not reasonable, while the flow factor is basically not limited upwards. This is because certain operation regions, such as the charging of the battery from a low SoC, require very small stoichiometric flow rates. Hence, it is possible to scale them with a very large flow factor without reaching the limiting pump capacity.

For every studied current density, the flow factor which yields the highest efficiency and the flow factor which yields the highest discharge capacity is identified using a parameter sweep. The optimization result is a look-up table consisting of tank SoC, charging/discharging current and associated flow factor, either to operate with highest efficiency or maximum discharge capacity. In practice, the BMS would select the appropriate flow factor from this look-up table.

8.5.2 Simulation results

As shown in Figure 8-4 a), amongst the displayed flow factors, a flow factor of three yields the highest RTSE for all studied currents. In fact, the optimal flow factor in terms of efficiency only varies little, as shown in Table 8-1.

For a smaller battery current, a larger flow factor decreases the discharge capacity due to additionally used SoC limitation and efficiency loss, as shown in Figure 8-4 b). Towards the nominal current, a substantially larger flow factor is reasonable. In fact, the optimal flow factors in terms of discharge capacity increase from 2.25 to 4.25 with an increasing current, as shown in Table 8-1.

For the conventional variable FRCS, the flow factor is allowed to vary with the current. Hence, for every simulated current, it is possible to select the flow factor that yields the highest RTSE or the largest discharge capacity. As shown in Table 8-1, the flow factors that yield the largest discharge capacity incline significantly faster with the current than the flow factors that yield the highest efficiency. This highlights the dilemma of deciding for highest efficiency or largest discharge capacity.

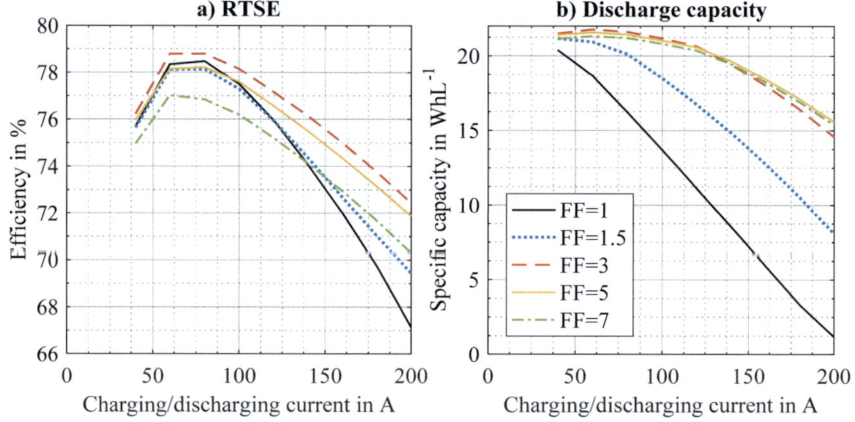

Figure 8-4: RTSE and specific discharge capacity for design 2.5 with the conventional variable FRCS using different flow factors (FF) in dependence of the applied current.

Table 8-1: Optimal flow factors for the conventional variable FRCS for different optimization objectives

Current in A	max. RTSE	max. Capacity
40	3.00	2.25
60	3.00	2.75
80	2.75	3.00
100	3.25	3.00
120	3.50	3.25
140	3.25	3.75
160	3.25	4.00
180	3.50	4.00
200	3.75	4.50

8.6 Variable flow rate - Innovative optimization approach

Both, constant and conventional variable FRCS have two major drawbacks. Experimental and/or model-based optimization of the flow factor requires conducting full cycles. This is very time consuming, in particular if we want to derive the optimal flow factor for a small current.

Further, the flow factor is always optimized for a complete charging/discharging cycle. Hence, it cannot be evaluated if the optimal flow factor at the beginning, in the middle or at the end of the charging or discharging process might vary.

Consequently, a new optimization approach that overcomes both hurdles is desirable. Therefore, the methodology previously presented by the author is applied and further developed [16].

8.6.1 Methodology

Using the instantaneous efficiency definition, presented previously in Section 3.4.3 on page 61, it is now possible to derive the system efficiency without conducting round-trips. Thus, the optimal flow rate for any operation point, defined by tank SoC and charging or discharging current, can be optimized individually.

Therefore, the model is adjusted in the following points. Tank SoC is now fixed at the operation point value and does no longer change during the simulation. The simulation lasts 10.000 seconds, to ensure that the system reaches its steady-state for every operation point.

To identify the optimal flow rate, MATLAB's Optimization Toolbox is deployed. The chosen solver, 'fminbnd' uses the golden-section search in combination with the successive parabolic interpolation to minimize a single-variable function [16, 99]. During the minimization, the variable is bounded by an upper and lower boundary.

Here, the single-variable function is a MATLAB function carrying out the operation point simulation with the Simulink model. The function receives the flow rate as input parameter and returns the relative system losses, as defined in Eq. (8-7), as result. Hence, the algorithm is able to vary the flow rate in order to minimize the losses. The model computes the system efficiency, η_{Sys}, for a given operation point as shown in Eq. (3-5) on page 61.

$$P_{\text{Loss,relative}} = 1 - \eta_{\text{Sys}} \tag{8-7}$$

The optimization process of an operation point is stopped, if the optimal flow rate is determined with a precision of 0.1 Lmin^{-1}.

It seems reasonable to select the nominal pump capacity as upper boundary for the optimization algorithm. However, it is found that this hinders the identification of the optimal flow rate if the optimum is close to the pump capacity. This happens for charging operations at a high SoC and for discharging operations at a low SoC. Hence, an upper limit of 200 Lmin^{-1} is used. This provides the optimization algorithm with sufficient buffer for any of the used pump capacities to properly determine the optimal flow rate in any operation point. If the optimization yields a flow rate that is larger than the pump capacity, it is set back to the pump capacity afterwards.

The lower boundary for the flow rate applied by the optimization algorithm has to satisfy two criteria:

1. It has to be equal to or larger than the minimally required flow rate according to Faraday's first law of electrolysis, as shown in Eq. (8-3).

2. The equation of the concentration overpotential contains a singularity. The argument of the logarithmic term may not become negative, as shown in Eq. (8-8) for charging the negative half-cell. Hence, the condition as shown in Eq. (8-9) can be derived. Herein, the mass transfer coefficient is related to the volumetric flow rate. From this, we can derive the minimally required flow rate to prevent the argument of the logarithm from becoming zero, as shown in Eq. (8-10). Again, this is illustrated for charging the negative half-cell. The negative half-cell represents the stronger condition due to its lower mass transfer coefficient.

The larger value of both conditions, complying with Faraday's law and delivering a positive argument for the logarithmic term of the concentration overpotential model, is deployed as the lower boundary for the optimization algorithm.

$$\left(1 - \frac{I_C}{2.38 A_E F k_{MT-} \, c_{3C-}}\right) > 0 \tag{8-8}$$

$$\frac{I_C}{2.38 A_E F c_{3C-}} < k_{MT-} = 1.608 \cdot 10^{-4} \left(\frac{Q_C}{CSA_E}\right)^{0.4} \tag{8-9}$$

$$Q_C > CSA_E \left(\frac{I_C}{2.38 A_E \cdot 1.608 \cdot 10^{-4} \cdot F c_{3C-}}\right)^{2.5} \tag{8-10}$$

The optimization of a sample operation point is shown in Figure 8-5. The algorithm requires eight iterations to determine the optimal flow rate for a tank SoC of 50 % and a charging current of 50 A to be 16.8 Lmin^{-1}. In this tank SoC, the system efficiency for charging the battery with 50 A is 86.4 %.

For the innovative variable FRCS, the optimal flow rate is determined for the following set of operation points. The tank SoC varies between 5 % and 95 % in steps of 1 %. The current varies between 40 and 200 A, which corresponds to a current density of 20 and 100 mAcm^{-2}, respectively. The current is varied in steps of 20 A, for both, the charging and discharging operation. Hence, a total number of 1,638 operation has to be optimized. As the optimization processes of different operation points are independent, they can be parallelized, e.g., by using MATLAB's Parallel Computing Toolbox. Also, the Simulink model can be set up in a manner that the flow rate is a so-called 'tunable parameter'. This allows for using the so-called 'Fast Restart' simulation mode. In this mode, the model remains compiled and the simulation starts substantially faster.

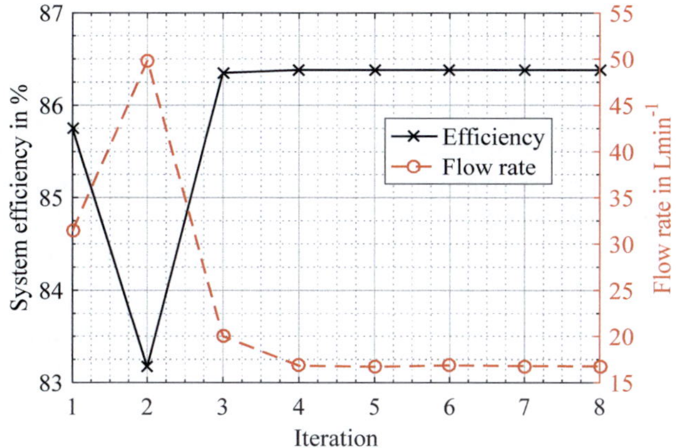

Figure 8-5 : Flow rate optimization of a sample operation point
with a tank SoC 50 % and a charging current of 50 A.

8.6.2   Simulation results

The optimization returns a three-dimensional look-up table, visualized in Figure 8-6 b).
For comparison, the corresponding flow rate computed with the conventional variable
FRCS with a flow factor of three is shown as well.

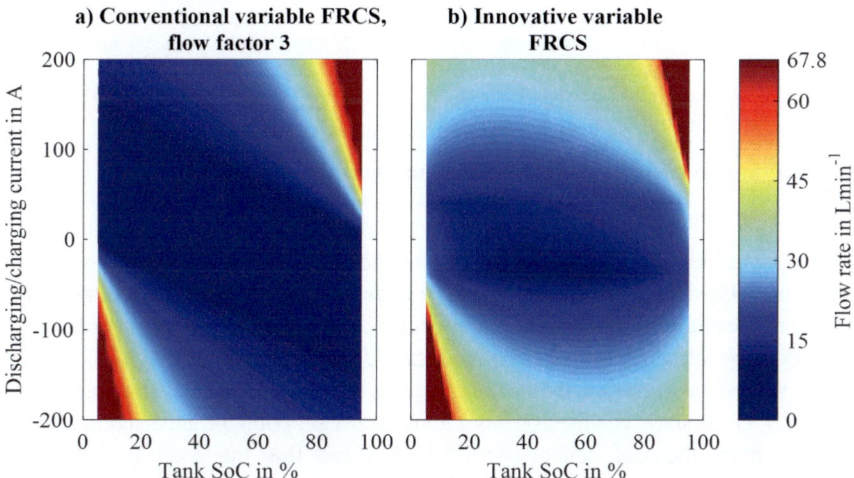

Figure 8-6: Comparison between the conventional and the innovative variable FRCS for the
test system with cell design 2.5.

In extreme operation points, namely charging the battery in a high SoC and discharging
the battery in a low SoC, the two FRCSs deviate less from each other, compared to the
more moderate operation points. In the latter, the innovative variable FRCS applies a
substantially larger flow rate than the conventional one.

Compared to the conventionally derived variable flow rates as shown in Figure 8-1 on page 121, the innovative FRCS shows a significantly deviating trend, as shown in Figure 8-7. The trend of a high flow rate at a high tank SoC for the charging operation and the trend of a high flow rate at a low tank SoC for the discharging operation is consistent. However, the optimized flow rates do neither show the monotonously increasing trend in the charging operation, nor the monotonously decreasing trend in the discharging operation towards a higher SoC.

To sum up, the innovative variable FRCS applies a higher flow rate in most of the operation points. In addition, it shows a completely different dependence on the SoC.

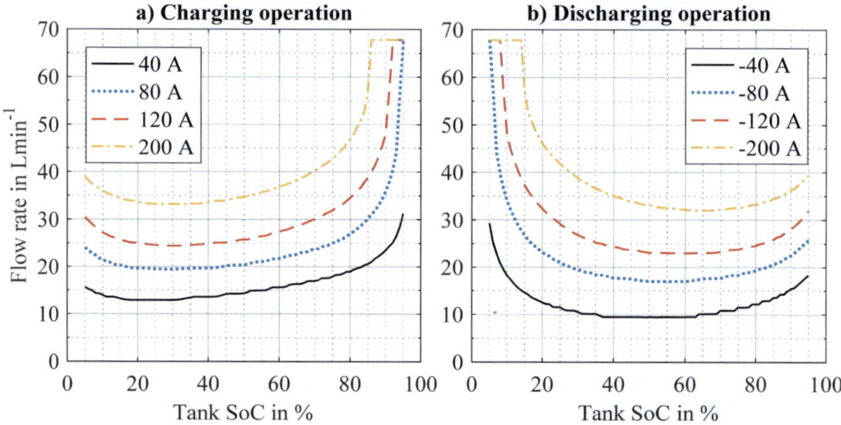

Figure 8-7: Optimal flow rate in dependence of tank SoC and current for design 2.5, derived with the innovative optimization approach.

8.7 Voltage-dependent FRCS – the stack voltage controller

8.7.1 Methodology

Besides its influence on the system efficiency, the FRCS also has a significant impact on the available discharge capacity, if cell or stack voltage limits are used to determine the end of the charging and discharging processes. Thus, the innovative variable FRCS is extended by introducing a superimposed stack voltage controller (VC).

During the normal operation, the stack is supplied with the optimal flow rate in terms of efficiency. If the stack voltage exceeds a given upper set point, or undershoots a given lower set point, the stack voltage controller takes over flow rate control, as shown in Figure 8-8.

The controller set points have to differ from the voltage limits used in the BMS for ending the charging or discharging process, respectively. If the controller is not activated until the absolute voltage limits are violated, the BMS will stop the charging or discharging process, respectively, before the controller has a chance to act on the stack voltage.

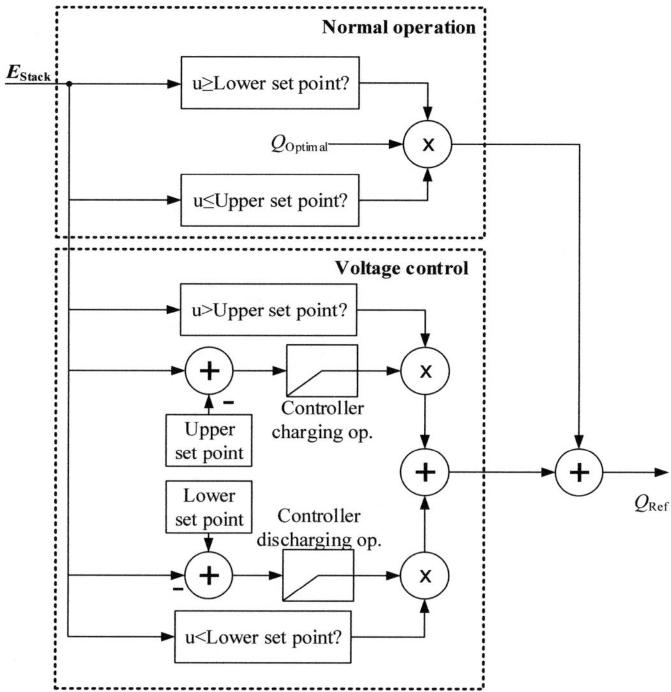

Figure 8-8: Structure of the proposed superimposed stack voltage controller

In this work, the controller is activated, if the stack voltage exceeds 65.6 V during a charging process or falls below 44.4 V during a discharging process. It then tries to maintain these voltages as long as possible by increasing the flow rate. Both set points correspond to a 400-mV buffer to the respective absolute voltage limits (10 mV per cell).

The gains for the proportional and integral parts of the controller are equal for both the controllers for the charging and discharging operation.

For a 40-cell stack, a proportional gain of $100\ \text{L(Vs)}^{-1}$ and an integral gain of $13.3\ \text{L(Vs}^2)^{-1}$ is heuristically determined. To illustrate these gains, let us consider the following example. If the stack voltage limit is exceeded by 20 mV (0.5 mV per cell) during the charging process, the proportional part of the controller instantly applies a flow rate of $120\ \text{Lmin}^{-1}$. If this violation lasts for 10 seconds, the integral part of the controller adds another $160\ \text{Lmin}^{-1}$. With these controller parameters, the controller quickly reacts on small violations of the upper and lower controller set point. Hence, it prevents the violation of the absolute voltage limits as long as the pump capacity is not fully exploited.

During the normal operation, the controller integrator is permanently reset. It receives the currently applied flow rate as an initialization value. This guarantees for a smooth transition between normal operation and voltage control mode.

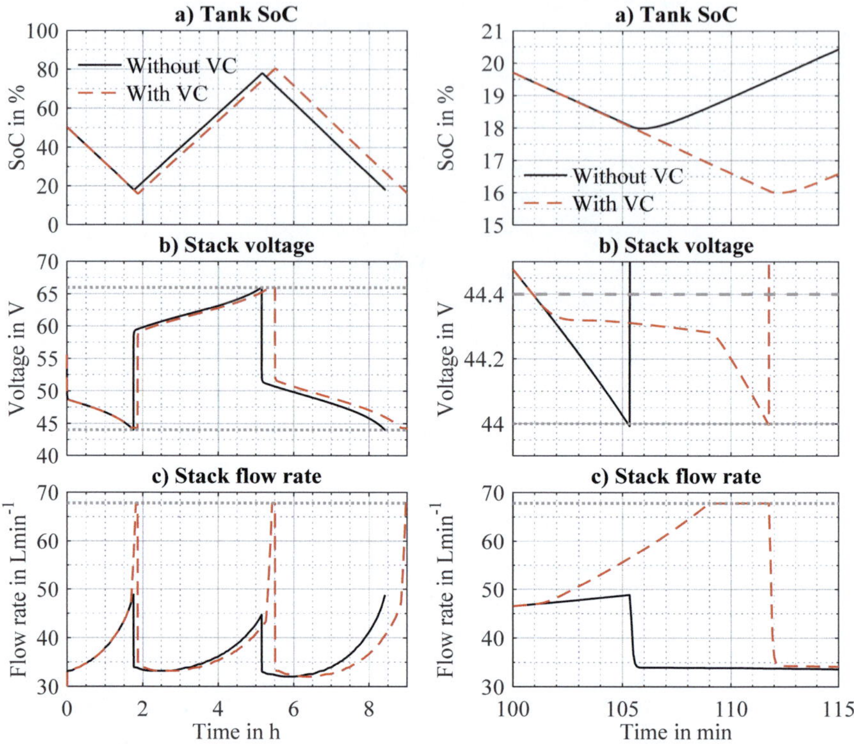

Figure 8-9: Demonstrating the stack voltage controller

Figure 8-10: Zoom into the demonstration of the stack voltage controller

To illustrate the function of the superimposed voltage controller, a charging/discharging cycle is simulated with the nominal current of 200 A for the innovative variable FRCS with and without the stack voltage controller, as show in Figure 8-9.

After 110.9 min, the stack voltage hits the lower set point during the pre-discharging cycle, as shown in Figure 8-10. Hence, in the system with voltage control, the stack voltage controller is activated. It quickly ramps up the flow rate to maintain the set-point of the stack voltage of 44.4 V (gray dashed line). Compared to the system without a voltage controller, the stack voltage controller significantly slows down the decay of the stack voltage as long as the nominal capacity of the pump is not fully exploited.

As a consequence, the lower voltage limit (gray dotted line) is reached 6.5 min later than without the controller. For the applied discharging current of 200 A and the 40-cell stack, this leads to an additional withdrawn electric charge of 867 Ah, which is an increase of 6.2 % referred to the total withdrawn charge during the pre-discharging process. While the system without the new controller only utilizes 72 % of the pump capacity, the stack voltage controller fully utilizes the capacity of 67.8 Lmin^{-1} (gray dotted line in Figure 8-10 c)).

8.8   Comparison of the FRCSs

8.8.1   Optimization objective highest RTSE

If optimized for obtaining the highest possible RTSE, the constant FRCS seems to outperform the other control strategies for a current larger than 60 A, as shown in Figure 8-11 a) and Table 8-2. However, this drastically reduces the discharge capacity, as shown in Figure 8-11 b) and Table 8-3. The constant FRCS loses 2.9 WhL^{-1} of nominal discharge capacity. This corresponds to a relative loss of 23 % and 20 % compared to the conventional and the innovative variable FRCS, respectively.

The innovative variable FRCS clearly outperforms the conventional one in terms of efficiency. The efficiency gain is up to 1.0 %-points but comes at the cost of a 0.7 WhL^{-1} decrease in the nominal discharge capacity, which is a relative loss of 4.5 %. Summarized, for currents larger than 60 A, the constant FRCS is the most efficient one, but faces a severe loss of nominal discharge capacity.

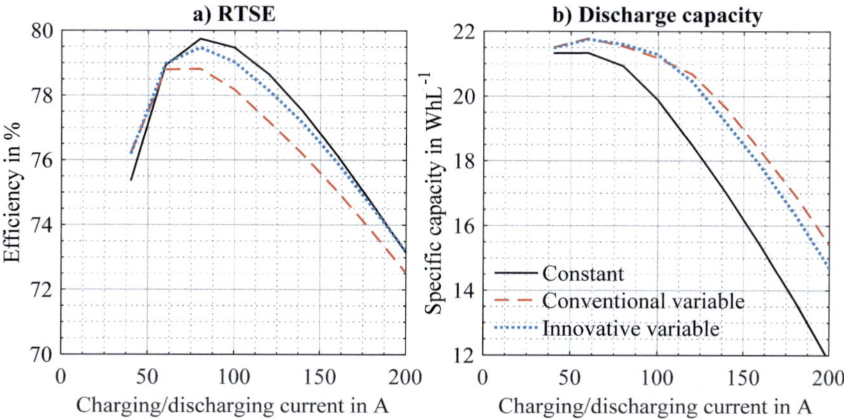

Figure 8-11: Comparison of the different FRCS with objective maximum system efficiency

Table 8-2: RTSE with FRCS optimized for highest RTSE

Current in A	Constant	Conv. variable	Innovative variable	$\eta_{Innovative}-\eta_{Constant}$	$\eta_{Innovative}-\eta_{Variable}$
40	75.4 %	76.2 %	76.2 %	0.8 %	0.0 %
60	78.9 %	78.8 %	79.0 %	0.1 %	0.2 %
80	79.7 %	78.8 %	79.5 %	-0.3 %	0.7 %
100	79.5 %	78.2 %	79.0 %	-0.4 %	0.8 %
120	78.7 %	77.2 %	78.2 %	-0.5 %	1.0 %
140	77.5 %	76.2 %	77.1 %	-0.4 %	1.0 %
160	76.1 %	75.0 %	75.9 %	-0.2 %	0.9 %
180	74.7 %	73.8 %	74.6 %	-0.1 %	0.8 %
200	73.2 %	72.5 %	73.2 %	0.0 %	0.6 %

Table 8-3: Specific discharge capacity in WhL^{-1} with FRCS optimized for highest RTSE

Current in A	Constant	Conv. variable	Innovative variable	$C_{Innovative}$ $-C_{Constant}$	$C_{Innovative}-$ $C_{Variable}$
40	21.3	21.5	21.5	0.2	0.0
60	21.3	21.8	21.8	0.4	0.0
80	20.9	21.5	21.6	0.7	0.1
100	19.9	21.2	21.3	1.4	0.1
120	18.5	20.7	20.4	2.0	-0.2
140	17.0	19.6	19.2	2.2	-0.5
160	15.4	18.3	17.8	2.4	-0.5
180	13.7	16.9	16.3	2.7	-0.6
200	11.8	15.4	14.7	2.9	-0.7

8.8.2 Optimization objective largest discharge capacity

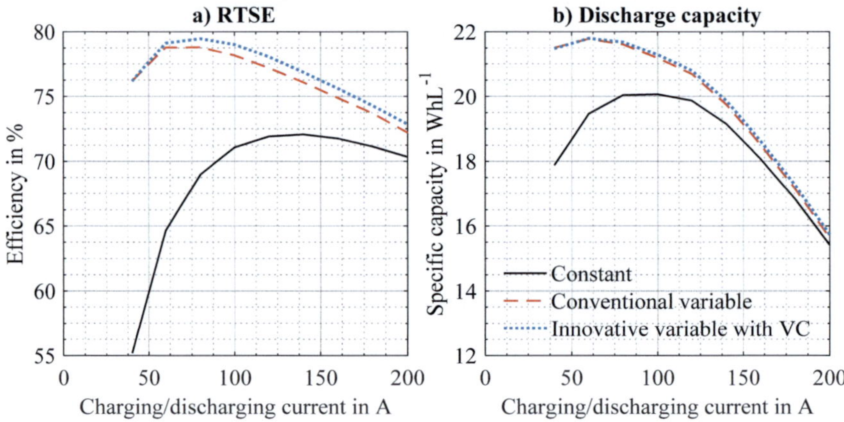

Figure 8-12: Comparison of the different FRCSs with objective maximum discharge capacity (VC = voltage controller)

For maximizing the discharge capacity, the innovative variable FRCS can be equipped with the proposed stack voltage controller. For the other two FRCSs, the flow factors which yield the maximum discharge capacity are deployed in this section.

Towards the nominal current, the constant FRCS yields a comparable discharge capacity as the two variable FRCSs, as shown in Figure 8-12 b) and in Table 8-5. The nominal discharge capacity is only 0.3 WhL^{-1} smaller than the nominal capacity of the innovative variable FRCS. However, towards a lower current, the gap in discharge capacity is substantially larger. For the lowest studied current, the gap is up to 3.6 WhL^{-1}. This corresponds to a relative capacity loss of 15 % compared to the two variable FRCSs.

In terms of efficiency, the large applied flow factor of the constant FRCS, which is unavoidable to obtain a comparable nominal discharge capacity, is very disadvantageous. With 2.5 %-points, the efficiency loss compared to the innovative variable FRCS is minimal for the nominal current density. However, for the lowest studied current density, the efficiency drops by as much as 21.0 %-points, as shown in Figure 8-12 a) and in Table 8-4.

Compared to the conventional variable FRCS, the innovative one yields the same or even a slightly increased discharge capacity. However, the true advantage of the new method is identified on the efficiency site. The innovative variable FRCS yields up to 0.9 %-points additional system efficiency. Hence, it can be concluded that the innovative variable FRCS in combination with the stack voltage controller now yields a higher efficiency without any loss of discharge capacity.

Table 8-4: RTSE with FRCS optimized for largest capacity

Current in A	Constant	Conv. variable	Innovative variable	$\eta_{Innovative}- \eta_{Constant}$	$\eta_{Innovative}- \eta_{Variable}$
40	55.2 %	76.2 %	76.2 %	21.0 %	0.0 %
60	64.7 %	78.8 %	79.1 %	14.5 %	0.3 %
80	69.0 %	78.8 %	79.5 %	10.5 %	0.7 %
100	71.1 %	78.2 %	79.0 %	7.9 %	0.9 %
120	71.9 %	77.2 %	78.0 %	6.1 %	0.9 %
140	72.1 %	76.1 %	76.9 %	4.8 %	0.8 %
160	71.8 %	74.9 %	75.6 %	3.9 %	0.7 %
180	71.1 %	73.7 %	74.3 %	3.1 %	0.6 %
200	70.3 %	72.2 %	72.9 %	2.5 %	0.7 %

Table 8-5: Specific discharge capacity in WhL^{-1} with FRCS optimized for largest capacity

Current in A	Constant	Conv. variable	Innovative variable	$C_{Innovative}- C_{Constant}$	$C_{Innovative}- C_{Variable}$
40	17.9	21.5	21.5	3.6	0.0
60	19.5	21.8	21.8	2.3	0.0
80	20.0	21.6	21.7	1.6	0.1
100	20.1	21.2	21.3	1.2	0.1
120	19.9	20.7	20.8	0.9	0.1
140	19.2	19.7	19.9	0.7	0.1
160	18.1	18.5	18.6	0.5	0.1
180	16.8	17.2	17.3	0.4	0.1
200	15.4	15.6	15.7	0.3	0.1

8.8.3 Two sample cycles

Two sample cycles are simulated with all studied FRCSs in order to demonstrate the advantage of the proposed innovative approach, combined with the stack voltage controller. The first sample cycle, shown in Figure 8-13, is conducted using a current of ±100 A and SoC limits of 10 % and 90 %. The second sample cycle, shown in Figure 8-14, is conducted with ±200 A between SoC limits of 20 % and 80 %. Note that SoC instead of voltage limits are deployed in this section to determine the end of the charging and discharging processes. Using SoC limits, the chronological sequences of the studied quantities show a better overlap which enables a better comparison of the different FRCSs.

For the constant FRCS, a flow factor of 4.35 is deployed for both cycles. For the conventional variable FRCS, a flow factor of three is applied for the first cycle and a flow factor of 4.5 is applied for the second cycle, in compliance with Table 8-1 on page 126. The innovative variable FRCS embraces the proposed superimposed stack voltage controller.

For both cycles, the SoC evolves identically for the three different FRCSs, as shown in Figure 8-13 a) and Figure 8-14 a). Due to the large applied flow rate, the constant FRCS yields the lowest charging and the highest discharging voltage, as shown in Figure 8-13 b) and Figure 8-14 b). Consequentially, this control strategy obtains lowest losses due to concentration overpotential, as shown in Figure 8-13 d) and Figure 8-14 d). The constant FRCS causes a constant pump power demand of 386 W over the whole operation time, as shown in Figure 8-13 e) and Figure 8-14 e).

For the predominant time period, the conventional variable FRCS applies the lowest flow rate. This results in the lowest pump power demand, but causes the highest concentration over-potential losses. At the end of the charging process with 200 A, the two variable FRCSs apply the same flow rate than the constant one, as shown in Figure 8-14 e). However, for the variable FRCSs the pump power demand for this flow rate is 942 W and thus 144 % larger than for the constant FRCS. The reason for that can be found in the pump efficiency. For the constant FRCS, the pump is assumed to yield a constant efficiency of 34.6 %. The variable speed pumps deployed for the variable FRCSs yield an efficiency of 14.1 %, if their capacity is fully exploited. The fixed speed pump is thus able to deliver the nominal flow rate more efficiently than the variable speed pumps. Unfortunately, the constant FRCS can only make use of this advantage for a very short period of time.

The innovative variable FRCS applies a larger flow rate than the conventional variable FRCS for the predominant part of the cycle time, as shown in Figure 8-13 c) and Figure 8-14 c). In the corresponding time, the innovative FRCS causes a higher pump power demand, as shown in Figure 8-13 e) and Figure 8-14 e).

The total losses clearly show that the constant FRCS is not a viable option for operating the battery with a current of 100 A, as shown in Figure 8-13 f). The strategy constantly causes additional losses of approximately 500 W over the complete cycle. For the operation with nominal current, the constant FRCS becomes more competitive, as shown in Figure 8-14 f). However, for the predominant part of the cycle time, it again causes the highest losses.

Amongst the two variable FRCSs, the innovative one yields the lowest losses during the two sample cycles. Hence, the claim of the optimization approach to yield the most efficient flow rate at any operation point it is clearly confirmed.

The total losses for both cycles increase towards a low SoC during the discharging process and towards a high SoC during the charging process, independently from the FRCS.

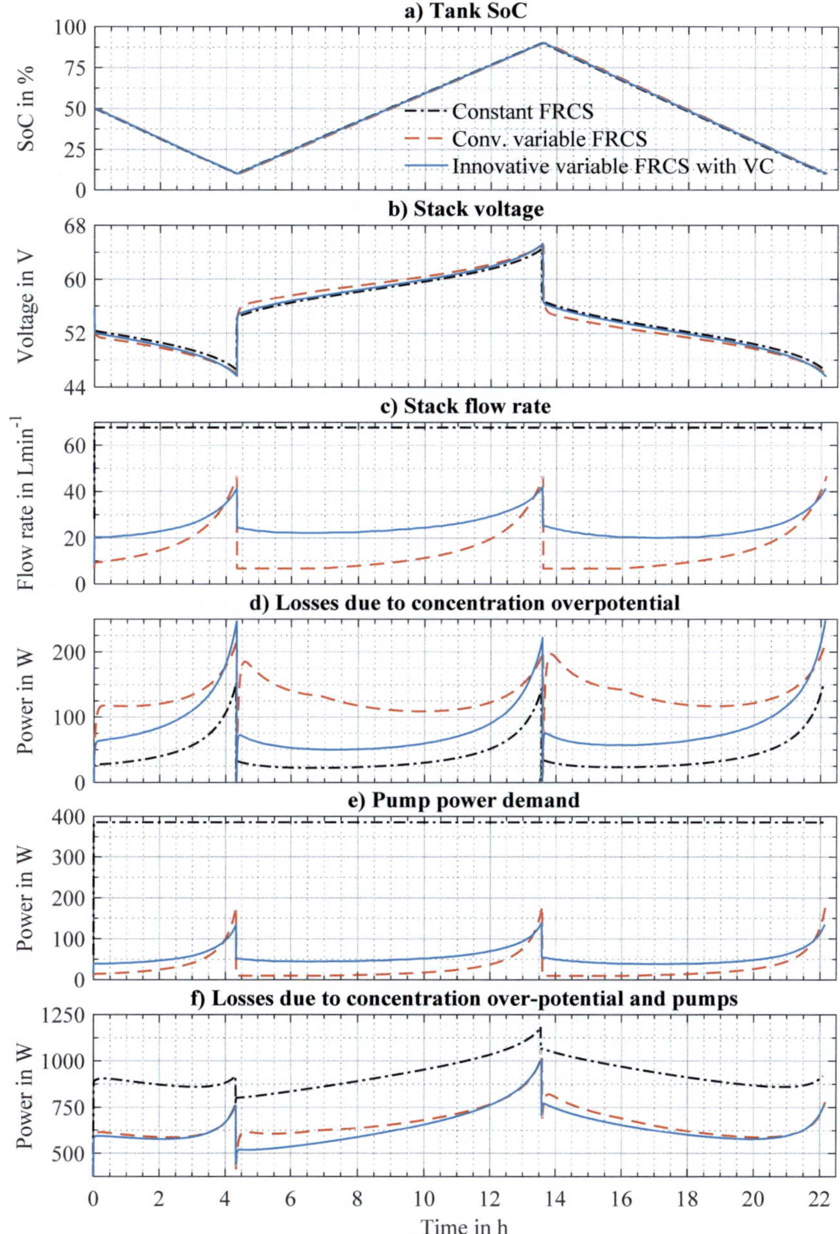

Figure 8-13: Tank SoC, stack voltage, flow rate and losses for a cycle with ±100 A and different FRCSs (VC = voltage controller)

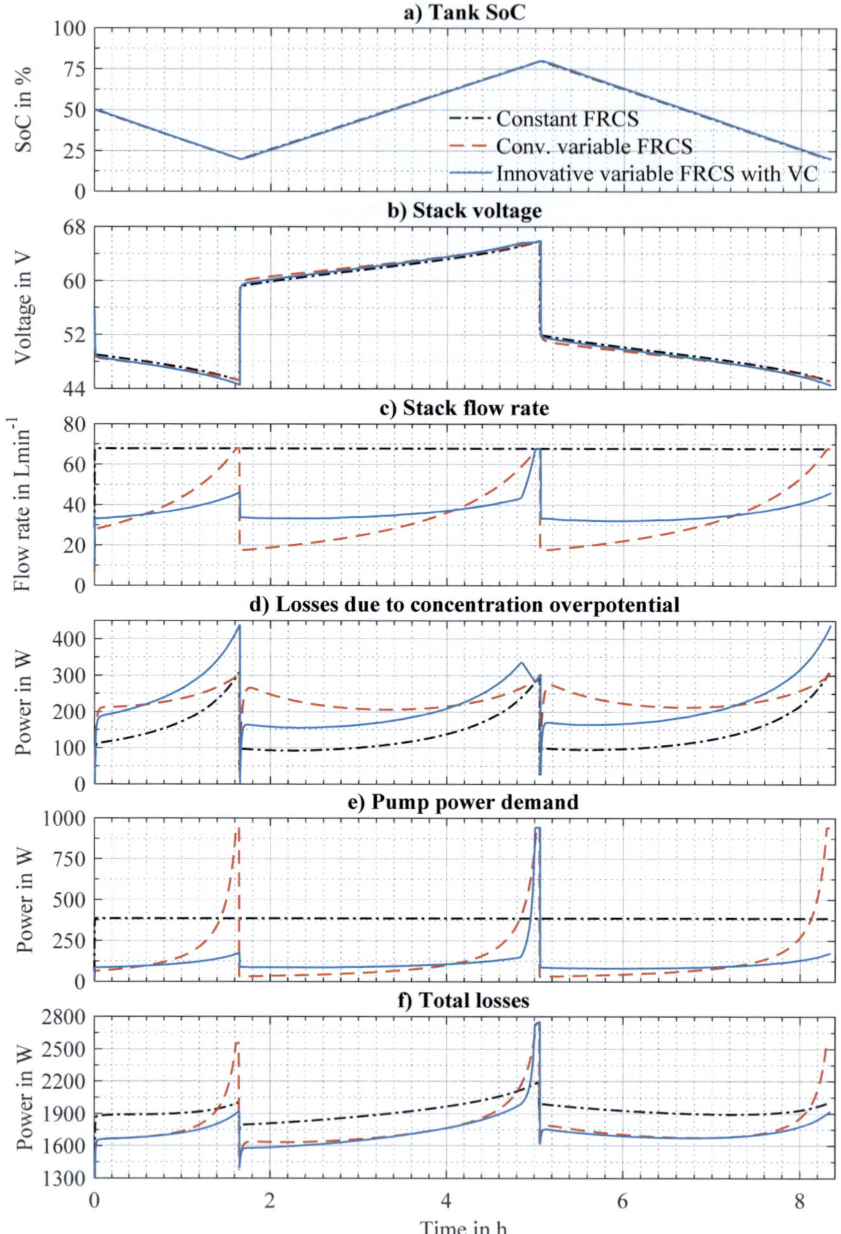

Figure 8-14: Tank SoC, stack voltage, flow rate and losses for a cycle with ±200 A and different FRCSs (VC = voltage controller)

8.9 Summary

8.9.1 Constant FRCS

The constant FRCS with a flow factor of 1.5 referred to the stoichiometric flow rate, required for charging the battery with the nominal current up to a tank SoC of 80 %, is the most efficient of the studied FRCSs. The reason is that the used SoC range and thus the yielded discharge capacity is reduced for this FRCS. As shown exemplarily in Figure 8-13 and Figure 8-14, the discharging operation in a low SoC and the charging operation in a high SoC causes the highest losses. Hence, if these SoC ranges are avoided, the efficiency rises. Unfortunately, limiting the SoC range leads to a reduced discharge capacity. Consequently, more electrolyte needs to be deployed to obtain the same discharge capacity, which increases costs and space requirements of the system.

The constant FRCS' loss of discharge capacity can be drastically reduced by applying a larger flow factor. In this case, the high constant pump power demand substantially decreases the efficiency, in particular for smaller currents.

To sum up, the constant FRCS is an option if one of the two following conditions is fulfilled. If these conditions do not apply, the constant FRCS should be avoided.

- There are no space limitations and costs of the electrolyte have declined significantly.

Or

- If activated, the battery always operates with a current close to its nominal value.

8.9.2 Variable FRCS

The use of a variable FRCS substantially increases discharge capacity and system efficiency. For the conventional variable FRCS, the flow factor is extensively optimized with a fine resolution of 0.25. Nevertheless, the newly proposed innovative variable FRCS outperforms the conventional one.

If both FRCSs are optimized with the objective of highest RTSE, the innovative FRCS yields up to 1.0 %-points additional efficiency. For the nominal current, the efficiency gain is 0.6 %-points. Similar to the constant FRCS, the innovative FRCS alone suffers a small loss of discharge capacity compared to the conventional variable one. However, the capacity loss is substantially smaller than for the constant FRCS.

The capacity loss is addressed by proposing a superimposed stack voltage controller which allows for combining high efficiency and high discharge capacity. If this controller is used, the innovative variable FRCS yields the highest efficiency and the highest discharge capacity in any case. Thus, it clearly is the superior FRCS.

8.9.3 Losses of the variable FRCSs

When compared to the already optimized conventional variable FRCS, the innovative one yields lower energy losses for every studied current, as shown in Figure 8-15.

Apart from the flow rate related correlations, we can clearly recognize the shifting of loss shares from vanadium crossover and shunt current losses towards ohmic losses and losses due to concentration overpotential that takes place towards larger currents.

If we take a look at the loss mechanisms which are most significantly affected by the FRCS, the innovative variable FRCS reduces the relevant losses by up to 18 %, namely for a battery current of 100 A and 120 A, as shown in Figure 8-16. For the nominal current, the reduction is 12 %.

Note that for a current of 40 A, the innovative variable FRCS yields higher losses due to concentration overpotential and pump power. However, if we take a look at the total losses, shown in Figure 8-15, we can see that the it still yields the lowest total losses. In this case, a part of the additionally deployed pump energy is obviously rewarded by a reduction of vanadium crossover. As the optimization algorithm targets on the reduction of total losses rather than of obviously flow rate related losses, it is also able to consider the influence of vanadium crossover.

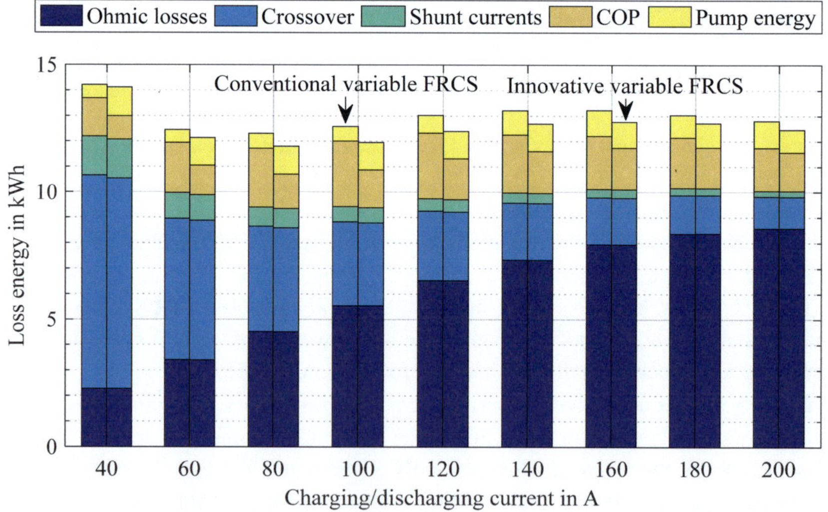

Figure 8-15: Loss distribution over applied current for the two variable FRCS strategies for cell design 2.5, optimized for capacity

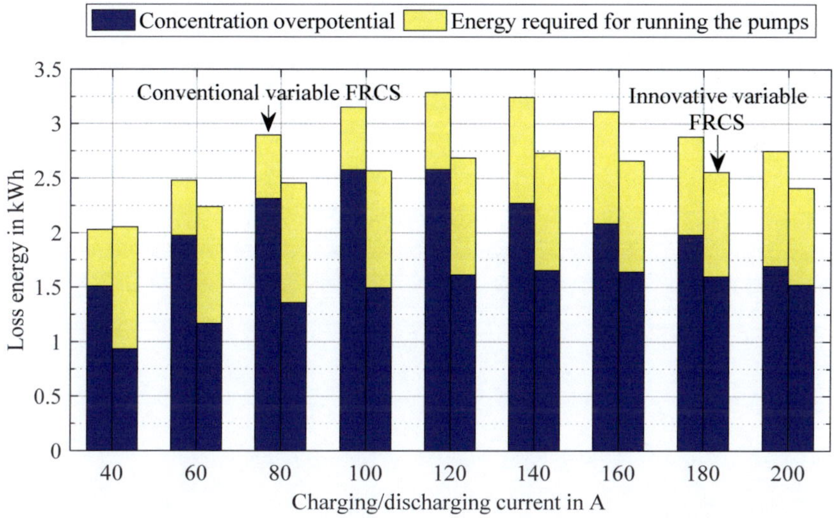

Figure 8-16: Losses due to concentration overpotential and pump power demand over applied current for the two variable FRCS strategies for cell design 2.5, optimized for capacity

Chapter 9

Summary and Outlook

9.1 Summary

This work consists of three parts. In the first part, a comprehensive, multi-physical lumped-parameter model is developed on the basis of numerous publications. The model is then validated using experimental data from three different manufacturers. However, because models typically found in the literature do not accurately reflect the behavior of the three real-life systems, an adaption method is introduced. Using the experimental data, the surface area for calculating the current density in the diffusion layer around the fibers of the graphite felt electrode is approximated. After its adaption, the model simulates the behavior of two systems very well; the simulation of the third system, although not ideal, is still more accurate than with the non-adapted literature approach.

In the second part of this work, the validated model is applied to the most comprehensive design study of a VRFB published to date. Twenty-four cell designs are evaluated for their eligibility for a single-stack system and a three-stack string system. In each system, 40-cell stacks are assembled virtually using the twenty-four designs. The design study presents a straightforward approach that uses the desired operational limits for cell voltage, SoC and flow rate to obtain a realistic system design. A unique feature of this design study is its simultaneous consideration of design and operational parameters. For each design, of both the single-stack and the three-stack string systems, the flow rate is optimized, to maximize the comparability of the designs.

The design study produced the following four key results:

- Increasing the electrode area and using a longer and narrower channel limit the impact of the shunt currents equally well. However, both measures face a strong saturation effect.

- For a given flow factor and aspect ratio, a larger electrode yields a larger electrolyte fluid velocity and thus a larger mass transfer coefficient. Hence, a larger electrode requires a lower flow factor for optimal operation.

- The large electrolyte demand of a large electrode causes an over-proportionally high pressure drop in the inlet and outlet channels.

- The electric series connection of three stacks facilitates the grid connection of the battery but reduces system efficiency by 1.6 to 3.3 %-points, depending on the applied current density. Only for higher current densities, however, it appears that the more efficient PCS resulting from higher input DC voltage might more than compensate for these efficiency losses.

For a single-stack system operating under variable load, a 2000-cm^2 cell with a long, medium-width channel is the best design. However, VRFB systems having comparable efficiencies can be built using electrode areas between 1000 and 4000 cm^2.

In a three-stack string, the negative effect of the shunt currents is strongly amplified. The advantage of a large electrode area is partly negated by larger diameters required for the external hydraulic circuit. Nevertheless, combining the largest electrode with the longest, narrowest channel yields the highest system efficiency for the three-stack string system. For such a system, small electrode areas and short, wide channels should be avoided. Cell design influences system efficiency much more strongly in multi-stack string systems than in single-stack systems.

In the third and final part, the novel flow rate control strategy (FRCS), published previously by the author, is presented and extended. The extension involves a stack voltage controller that promotes efficient operation and good exploitation of the electrolyte's energy density.

The constant FRCS appears to be an option for maximizing the VRFB's efficiency. The downside, however, is a massive loss of discharge capacity. Compared to the conventional variable FRCS, the novel version is up to 1.0 %-points more efficient.

If the goal is to maximize the VRFB's discharge capacity, the constant FRCS is competitive. However, for lower current densities, system efficiency drops by as much as 21.0 %-points compared to the variable FRCSs. For all current densities, the conventional variable FRCS yields the same discharge capacities as the extended novel version using the stack voltage controller. However, the latter increases system efficiency by up to 0.9 %-points. In summary, when combined with the simple but effective stack voltage controller, the extended novel FRCS is the superior flow rate control strategy for VRFBs.

9.2 Outlook

9.2.1 General outlook for VRFB research

Impact of design and operational parameters on the efficiency of a VRFB is limited. Model-based approaches can hardly decrease the two major loss shares ohmic losses and vanadium crossover. However, it is of outstanding importance to reduce these losses, e.g., by reducing the contact resistances of the different flow cell components. Also, material research should be intensively conducted to produce a membrane with an increased selectivity, but a comparable or even higher ionic conductance. As shown in Figure 9-1, a reduced cell resistance in particular improves the efficiency for larger currents. Decreased diffusion coefficients of the vanadium ions in the membrane promotes significantly higher partial load efficiency. Combining both the smaller resistance and the smaller diffusion coefficients increases system efficiency by up to 9 %-points over the whole operational area, compared to today's systems.

Naturally, this significantly increases the available discharge capacity also, as shown in Figure 9-1 b). A decreased cell resistance increases the discharge capacity more strongly than decreased diffusion coefficients. Finally, a reduced resistance also enables higher current densities, resulting in an increasing power density. This subsequently reduces the power-related system cost, which is urgently required.

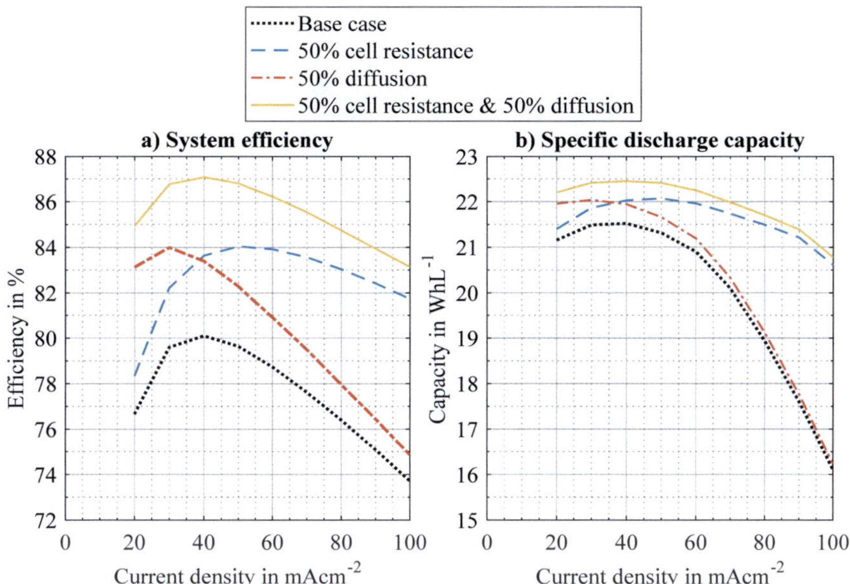

Figure 9-1: Reduced cell resistance and diffusion coefficients increase the round-trip efficiency and the discharge capacity of future VRFBs.

9.2.2 Outlook on model-based works

In the large number of published model-based studies, the model validation is not a very popular topic, in particular regarding commercial-scale systems. Hence, the collaboration of manufacturers and research institutes should be intensified, in order to make sure that the presented virtual models also unfold an actual practical benefit.

Regarding the model extensions, the plug flow reactor principle can overcome the simplification of a perfectly mixed cell [19, 100]. Further, first works try to capture the topic of ageing [27]. For long-term operation, the effect of bulk electrolyte and water transfer across the membrane should be included in lumped-parameter models, e.g. as shown in [24].

The standard electrolyte is an oversaturated solution. Thus, a certain temperature window has to be kept in order to prevent the precipitation of the vanadium salts. Consequentially, sophisticated thermal models of the VRFB are of high interest and should be included in future studies.

Appendix A

Summary of the essential equations used in the model

A.1 State-of-charge

$$SoC_{\mathrm{T}} = \frac{\sqrt{\dfrac{c_{2\mathrm{T}-} \cdot c_{5\mathrm{T}+}}{c_{3\mathrm{T}-} \cdot c_{4\mathrm{T}+}}}}{1 + \sqrt{\dfrac{c_{2\mathrm{T}-} \cdot c_{5\mathrm{T}+}}{c_{3\mathrm{T}-} \cdot c_{4\mathrm{T}+}}}}$$

A.2 Tank concentration of ionic species

$$V_{\mathrm{T}}\,\frac{\mathrm{d}}{\mathrm{d}t}c_{i\mathrm{T}-} = J^{\mathrm{out}}_{i\mathrm{Stacks}-} - Q_{\mathrm{T}}c_{i\mathrm{T}-},\ i\in\{2,3,4,5\}$$

$$V_{\mathrm{T}}\,\frac{\mathrm{d}}{\mathrm{d}t}c_{i\mathrm{T}+} = J^{\mathrm{out}}_{i\mathrm{Stacks}+} - Q_{\mathrm{T}}c_{i\mathrm{T}+},\ i\in\{2,3,4,5\}$$

$$J^{\mathrm{out}}_{i\mathrm{Stacks}-} = \sum_{m=1}^{N_{\mathrm{S}}} \frac{Q_{\mathrm{S}m}}{N_{\mathrm{C}}} \sum_{n=1}^{N_{\mathrm{C}}} c^{\mathrm{out}}_{i\mathrm{C}mn-}$$

$$J^{\mathrm{out}}_{i\mathrm{Stacks}+} = \sum_{m=1}^{N_{\mathrm{S}}} \frac{Q_{\mathrm{S}m}}{N_{\mathrm{C}}} \sum_{n=1}^{N_{\mathrm{C}}} c^{\mathrm{out}}_{i\mathrm{C}mn+}$$

A.3 Cell concentration of ionic species

$$V_{\mathrm{E}}\frac{\mathrm{d}}{\mathrm{d}t}\begin{pmatrix} c^{\mathrm{out}}_{2\mathrm{C}-} \\ c^{\mathrm{out}}_{3\mathrm{C}-} \\ c^{\mathrm{out}}_{4\mathrm{C}-} \\ c^{\mathrm{out}}_{5\mathrm{C}-} \end{pmatrix} = \frac{1}{\mathrm{F}}\begin{pmatrix} +I_{\mathrm{C}} \\ -I_{\mathrm{C}} \\ 0 \\ 0 \end{pmatrix} + \begin{pmatrix} J_{2\mathrm{Cross}-} \\ J_{3\mathrm{Cross}-} \\ J_{4\mathrm{Cross}-} \\ J_{5\mathrm{Cross}-} \end{pmatrix} + \frac{Q_{\mathrm{S}}}{N_{\mathrm{C}}}\begin{pmatrix} c_{2\mathrm{T}-} - c^{\mathrm{out}}_{2\mathrm{C}-} \\ c_{3\mathrm{T}-} - c^{\mathrm{out}}_{3\mathrm{C}-} \\ c_{4\mathrm{T}-} - c^{\mathrm{out}}_{4\mathrm{C}-} \\ c_{5\mathrm{T}-} - c^{\mathrm{out}}_{5\mathrm{C}-} \end{pmatrix}$$

$$V_{\mathrm{E}}\frac{\mathrm{d}}{\mathrm{d}t}\begin{pmatrix} c^{\mathrm{out}}_{2\mathrm{C}+} \\ c^{\mathrm{out}}_{3\mathrm{C}+} \\ c^{\mathrm{out}}_{4\mathrm{C}+} \\ c^{\mathrm{out}}_{5\mathrm{C}+} \end{pmatrix} = \frac{1}{\mathrm{F}}\begin{pmatrix} 0 \\ 0 \\ -I_{\mathrm{C}} \\ +I_{\mathrm{C}} \end{pmatrix} + \begin{pmatrix} J_{2\mathrm{Cross}+} \\ J_{3\mathrm{Cross}+} \\ J_{4\mathrm{Cross}+} \\ J_{5\mathrm{Cross}+} \end{pmatrix} + \frac{Q_{\mathrm{S}}}{N_{\mathrm{C}}}\begin{pmatrix} c_{2\mathrm{T}+} - c^{\mathrm{out}}_{2\mathrm{C}+} \\ c_{3\mathrm{T}+} - c^{\mathrm{out}}_{3\mathrm{C}+} \\ c_{4\mathrm{T}+} - c^{\mathrm{out}}_{4\mathrm{C}+} \\ c_{5\mathrm{T}+} - c^{\mathrm{out}}_{5\mathrm{C}+} \end{pmatrix}$$

$$\begin{pmatrix} J_{2\mathrm{Cross}-} \\ J_{3\mathrm{Cross}-} \\ J_{4\mathrm{Cross}-} \\ J_{5\mathrm{Cross}-} \end{pmatrix} = \frac{A_{\mathrm{Mem}}}{\delta_{\mathrm{Mem}}}\begin{pmatrix} (c_{2\mathrm{C}+} - c_{2\mathrm{C}-})D_{2\mathrm{Mem}} - (c_{4\mathrm{C}+} - c_{4\mathrm{C}-})D_{4\mathrm{Mem}} - 2(c_{5\mathrm{C}+} - c_{5\mathrm{C}-})D_{5\mathrm{Mem}} \\ (c_{3\mathrm{C}+} - c_{3\mathrm{C}-})D_{3\mathrm{Mem}} + 2(c_{4\mathrm{C}+} - c_{4\mathrm{C}-})D_{4\mathrm{Mem}} + 3(c_{5\mathrm{C}+} - c_{5\mathrm{C}-})D_{5\mathrm{Mem}} \\ 0 \\ 0 \end{pmatrix}$$

$$\begin{pmatrix} J_{2\mathrm{Cross}+} \\ J_{3\mathrm{Cross}+} \\ J_{4\mathrm{Cross}+} \\ J_{5\mathrm{Cross}+} \end{pmatrix} = \frac{A_{\mathrm{Mem}}}{\delta_{\mathrm{Mem}}}\begin{pmatrix} 0 \\ 0 \\ 3(c_{2\mathrm{C}-} - c_{2\mathrm{C}+})D_{2\mathrm{Mem}} + 2(c_{3\mathrm{C}-} - c_{3\mathrm{C}+})D_{3\mathrm{Mem}} + (c_{4\mathrm{C}-} - c_{4\mathrm{C}+})D_{4\mathrm{Mem}} \\ -2(c_{2\mathrm{C}-} - c_{2\mathrm{C}+})D_{2\mathrm{Mem}} - (c_{3\mathrm{C}-} - c_{3\mathrm{C}+})D_{3\mathrm{Mem}} + (c_{5\mathrm{C}-} - c_{5\mathrm{C}+})D_{5\mathrm{Mem}} \end{pmatrix}$$

Appendix A – Summary of the essential equations used in the model

A.4 Open circuit voltage

$$E_{\text{OCVC}} = \widetilde{E}^0 + \frac{GT}{F} \ln\left(\frac{c_{2C-} \cdot c_{5C+}}{c_{3C-} \cdot c_{4C+}}\right)$$

A.5 Ohmic overpotential

$$E_{\text{Ohm}} = I_C R_C$$

A.6 Concentration overpotential

$$E_{\text{COPcharging}-} = -\frac{GT}{F} \ln\left(1 - \frac{|i_{\text{DL}}|}{F k_{\text{MT}-} \cdot c_{3C-}}\right), \text{ for } \frac{|i_{\text{DL}}|}{F k_{\text{MT}-} \cdot c_{3C-}} < 1$$

$$E_{\text{COPdischarging}-} = -\frac{GT}{F} \ln\left(1 - \frac{|i_{\text{DL}}|}{F k_{\text{MT}-} \cdot c_{2C-}}\right), \text{ for } \frac{|i_{\text{DL}}|}{F k_{\text{MT}-} \cdot c_{2C-}} < 1$$

$$E_{\text{COPcharging}+} = -\frac{GT}{F} \ln\left(1 - \frac{|i_{\text{DL}}|}{F k_{\text{MT}+} \cdot c_{4C+}}\right), \text{ for } \frac{|i_{\text{DL}}|}{F k_{\text{MT}+} \cdot c_{4C+}} < 1$$

$$E_{\text{COPdischarging}+} = -\frac{GT}{F} \ln\left(1 - \frac{|i_{\text{DL}}|}{F k_{\text{MT}+} \cdot c_{5C+}}\right), \text{ for } \frac{|i_{\text{DL}}|}{F k_{\text{MT}+} \cdot c_{5C+}} < 1$$

$$k_{\text{MT}-} = 1.608 \cdot 10^{-4} \left(\frac{Q_C}{CSA_E}\right)^{0.40}$$

$$k_{\text{MT}+} = 2.613 \cdot 10^{-4} \left(\frac{Q_C}{CSA_E}\right)^{0.40}$$

$$i_{\text{DL}} = \frac{I_C}{2.38 A_E}$$

A.7 Total cell voltage

$$E_C = E_{\text{OCVC}} + E_{\text{COPcharging}-} + E_{\text{COPcharging}+} + I_C R_C, \text{ charging process}$$

$$E_C = E_{\text{OCVC}} - E_{\text{COPdischarging}-} - E_{\text{COPdischarging}+} + I_C R_C, \text{ discharging process}$$

A.8 Pressure drop of the complete cell

$$\Delta p_C = \beta Q_C + \gamma Q_C^{\,2}$$

Appendix A – Summary of the essential equations used in the model

A.9 Pressure drop in pipes

$$\Delta p_P = 8f \frac{l_P}{d_P{}^5} \frac{\rho_{El}}{\pi^2} Q_P{}^2$$

$$f = \begin{cases} {}^{64}/_{Re_P}, & \text{for } Re_P \leq 2{,}300 \\[2mm] f_{La} + \dfrac{f_{Tu} - f_{La}}{2{,}300}(Re_P - 2{,}300), & \text{for } 2{,}300 < Re_P \leq 4{,}000 \\[2mm] \left(1.8 \log_{10}\left({}^{6.9}/_{Re_P} + \left(\dfrac{{}^{\epsilon}/_{d_P}}{3.7}\right)^{1.11}\right)\right)^{-2}, & \text{for } 4{,}000 \leq Re_P \end{cases}$$

$$Re_P = \frac{\rho_{El} v_P d_P}{\mu_{El}} = 4 \frac{\rho_{El} Q_P}{\mu_{El} \pi d_P}$$

A.10 Pressure drop due to orifices

$$\Delta p_{\text{Orifice}} = 8k_L \frac{\rho_{El}}{d_{\text{Orifice}}{}^4 \pi^2} Q_{\text{Orifice}}{}^2$$

A.11 Pump power

$$P_{\text{Pumps}} = 2 \frac{\Delta p_{\text{Total}} Q_{\text{Pump}}}{\eta_{\text{Pump}}(Q_{\text{Pump}})}$$

Appendix B

Additional data and calculations

B.1 Calculation of the tank SoC from the electrolyte OCV

In the model, the OCV is derived using the Nernst equation. With the assumption of a constant hydrogen proton concentration, Eq. (A-1) is derived.

$$E_{\mathrm{OCV}} = \tilde{E}^0 + \frac{GT}{F} \ln \left(\frac{c_{2\mathrm{T}-} \cdot c_{5\mathrm{T}+}}{c_{3\mathrm{T}-} \cdot c_{4\mathrm{T}+}} \right) \tag{A-1}$$

Assuming an ideally balanced electrolyte, we can calculate the identical SoCs of both electrolytes as shown in Eq. (A-2).

$$SoC_{\mathrm{T}} = \frac{c_{2\mathrm{T}-}}{c_{2\mathrm{T}-} + c_{3\mathrm{T}-}} = \frac{c_{5\mathrm{T}+}}{c_{4\mathrm{T}+} + c_{5\mathrm{T}+}} \tag{A-2}$$

The sum of the concentrations of V^{2+} and V^{3+} ions in the negative electrolyte and the sum of the concentrations of VO^{2+} and VO_2^+ ions in the positive electrolyte both equal the total vanadium concentration, c_V, as shown in Eq. (A-3).

$$c_{2\mathrm{T}-} + c_{3\mathrm{T}-} = c_{4\mathrm{T}+} + c_{5\mathrm{T}+} = c_V \tag{A-3}$$

Now we can express the concentrations exclusively by SoC and total vanadium concentration, as shown in Eq. (A-4).

$$\begin{pmatrix} c_{2\mathrm{T}-} \\ c_{3\mathrm{T}-} \\ c_{4\mathrm{T}+} \\ c_{5\mathrm{T}+} \end{pmatrix} = \begin{pmatrix} SoC_{\mathrm{T}} \\ 1 - SoC_{\mathrm{T}} \\ 1 - SoC_{\mathrm{T}} \\ SoC_{\mathrm{T}} \end{pmatrix} c_V \tag{A-4}$$

If we replace all concentrations in Eq. (A-1) by the expressions using the SoC, we can derive Eq. (A-5), which represents a common simplification for the OCV(SoC) relation.

$$\begin{aligned} E_{\mathrm{OCV}} &= \tilde{E}^0 + \frac{GT}{F} \ln \left(\frac{SoC_{\mathrm{T}} c_V SoC_{\mathrm{T}} c_V}{(1 - SoC_{\mathrm{T}}) c_V (1 - SoC_{\mathrm{T}}) c_V} \right) \\ &= \tilde{E}^0 + \frac{GT}{F} \ln \left(\frac{SoC_{\mathrm{T}}^2}{(1 - SoC_{\mathrm{T}})^2} \right) \end{aligned} \tag{A-5}$$

The comparison of the argument of the logarithmic term in Eq. (A-1) with the one in Eq. (A-5) yields Eq. (A-6).

$$\frac{SoC_{\mathrm{T}}^2}{(1 - SoC_{\mathrm{T}})^2} = \frac{c_{2\mathrm{T}-} \cdot c_{5\mathrm{T}+}}{c_{3\mathrm{T}-} \cdot c_{4\mathrm{T}+}} \tag{A-6}$$

Finally, a SoC dependence on all four different vanadium ions can be derived, as shown in Eq. (A-7).

Appendix B – Additional data and calculations

$$SoC_T = \frac{\sqrt{\dfrac{c_{2T-} \cdot c_{5T+}}{c_{3T-} \cdot c_{4T+}}}}{1 + \sqrt{\dfrac{c_{2T-} \cdot c_{5T+}}{c_{3T-} \cdot c_{4T+}}}} \tag{A-7}$$

B.2 Constant parameters

Variable	Quantity	Value	Unit	Source
c_V	Total vanadium concentration	1,600	$molm^{-3}$	[101]
d_F	Fiber diameter of the graphite felt	17.6×10^{-6}	m	[47]
δ_{Mem}	Membrane thickness	127×10^{-6}	m	[28]
D_{2Mem}	Diffusion coefficient of V^{2+} ions in the membrane	8.8×10^{-12}	m^2s^{-1}	[102]
D_{3Mem}	Diffusion coefficient of V^{3+} ions in the membrane	3.2×10^{-12}	m^2s^{-1}	[102]
D_{4Mem}	Diffusion coefficient of VO^{2+} ions in the membrane	6.9×10^{-12}	m^2s^{-1}	[102]
D_{5Mem}	Diffusion coefficient of VO_2^+ ions in the membrane	5.8×10^{-12}	m^2s^{-1}	[102]
D_{2El}	Diffusion coefficient of V^{2+} ions in the electrolyte	2.4×10^{-10}	m^2s^{-1}	[55]
D_{3El}	Diffusion coefficient of V^{3+} ions in the electrolyte	2.4×10^{-10}	m^2s^{-1}	[55]
D_{4El}	Diffusion coefficient for VO^{2+} ions in the electrolyte	3.9×10^{-10}	m^2s^{-1}	[55]
D_{5El}	Diffusion coefficient for VO_2^+ ions in the electrolyte	3.9×10^{-10}	m^2s^{-1}	[55]
F	Faraday constant	96,485	$Asmol^{-1}$	–
G	Universal gas constant	8.314	$J(molK)^{-1}$	–
K_{KC}	Kozeny Carman constant	4.28	-	[67]
T	Temperature	298.15	K	–
ε	Graphite felt porosity	0.93	-	[47]
ϵ	Internal roughness height	1.5×10^{-6}	m	[72]
μ	Electrolyte dynamic viscosity	4.928×10^{-3}	Pas	[67]
ρ	Electrolyte density	1,354	kgm^{-3}	[103]
ψ	Area specific resistance	1.5×10^{-4}	Ωm^{-2}	Assumed

Appendix C

Nomenclature

C.1 Abbreviations

AC	Alternating current
ASR	Area specific resistance
BEP	Best efficiency point
BESS	Battery energy storage system
BMS	Battery management system
CFD	Computational fluid dynamics
COP	Concentration overpotential
DC	Direct current
DSM	Demand side management
EFR	Enhanced frequency regulation
EMF	Electromotive force
ESS	Energy storage system
FEA	Finite element analysis
FF	Flow factor
FRCS	Flow rate control strategy
NASA	National aeronautics and space administration
OCV	Open circuit voltage
PCS	Power conditioning system
PV	Photovoltaics
RES	Renewable energy sources
RTSE	Root mean square error
SAIDI	System average interruption duration index
SoC	State-of-charge
TSO	Transport system operator
VC	Voltage controller
VRFB	Vanadium Redox Flow Battery

C.2 Variables

a	Activity	α	Charge transfer coefficient
A	Area	β	Coefficient
c	Concentration	γ	Coefficient
C	Capacity	δ	Thickness
CSA	Cross-sectional-area	ε	Porosity
d	Diameter	η	Efficiency
D	Diffusion coefficient	κ	Permeability
E	Voltage	μ	Viscosity
f	Friction factor	ρ	Density
F	Faraday constant	σ	Conductivity
g	Gravity constant	ψ	Specific electric resistance
G	Universal gas constant		
Geo	Geometry factor		
h	Height		
i	Electric current density		
I	Electric current		
j	Ionic flux density		
J	Ionic flux		
k	Constant / Coefficient		
K	Constant		
l	Length		
n	Counting variable		
N	Number		
p	Pressure		
P	Power		
q	Quality criterion		
Q	Volumetric flow rate		
R	Resistance		
Re	Reynolds number		
$RMSE$	Root-mean-square-error		
$RTSE$	Round-trip system efficiency		
s	Specific surface area		
Sh	Sherwood number		
SoC	State-of-charge		
t	Time		
T	Temperature		
v	Velocity		
V	Volume		
w	Width		
W	Energy		
x	Location		

C.3 Subscripts and superscripts

Subscripts	
+	Positive half-side / half-cell / electrolyte
−	Negative half-side / half-cell / electrolyte
0	Initial
2	Vanadium V^{2+}
3	Vanadium V^{3+}
4	Vanadium VO^{2+}
5	Vanadium VO_2^+
Act	Activation
An	Anodic
Aux	Auxiliary
C	Cell
Ca	Cathodic
Ch	Channel
Col	Coulomb
COP	Concentration overpotential
Cross	Crossover and side-reactions
DC	Direct current
DL	Diffusion layer
eff	Effective
E	Electrode
El	Electrolyte
Ene	Energy
F	Fiber
FF	Flow factor
H	Hydrogen

i	Counting variable
j	Current density
k	Counting variable
KC	Kozeny-Carman
l	Counting variable
L	Loss
La	Laminar
m	Counting variable
M	Manifold
Mem	Membrane
MT	Mass transfer
n	Counting variable
Nom	Nominal
O	Object
OCV	Open-circuit-voltage
P	Pipe
PCS	Power conditioning system
RC	Rate constant
S	Stack
Sys	System
T	Tank
TF	Transfer
Tu	Turbulent
V	Vanadium
Vol	Voltage

Superscripts	
0	Standard
out	Output

Bibliography

[1] Hagedorn, N. H. 1984. *NASA Redox Storage System Development Project*. NASA.

[2] Bundesnetzagentur. 2016. *Quality of supply*. http://www.bundesnetzagentur.de/ cln_1432/EN/Areas/Energy/Companies/SecurityOfSupply/QualityOfSupply/ QualityOfSupply_node.html. Accessed 20 March 2017.

[3] EEX Transparency Platform. 2016. *Installed Storage Capacity*. https://www.eex-transparency.com/homepage/power/germany/storage/capacity/installed-capacity-. Accessed 20 March 2017.

[4] National Grid. 2016. *EFR Full Results*. http://www2.nationalgrid.com/WorkArea/ DownloadAsset.aspx?id=8589936484. Accessed 5 March 2017.

[5] Wilcox, G. J. 2015. LA County declares state of emergency over Porter Ranch gas leak. *Los Angeles Daily News* 2015 (Dec. 2015).

[6] Geuss, M. 2016. *SoCal utility will buy 80MWh of battery storage from Tesla after methane leak*. https://arstechnica.com/business/2016/09/socal-utility-will-buy-80mwh-of-battery-storage-from-tesla-after-methane-leak/. Accessed 20 March 2017.

[7] steag GmbH. 2017. *STEAG large-scale battery systems*. http://www.steag-grossbatterie-system.com/en/index.html. Accessed 20 March 2017.

[8] Noack, J., Wietschel, L., Roznyatovskaya, N., Pinkwart, K., and Tübke, J. 2016. Techno-Economic Modeling and Analysis of Redox Flow Battery Systems. *Energies* 9, 8, 627.

[9] König, S., Suriyah, M. R., and Leibfried, T. 2015. Model based examination on influence of stack series connection and pipe diameters on efficiency of vanadium redox flow batteries under consideration of shunt currents. *Journal of Power Sources* 281, 272–284.

[10] König, S., Wöber, A., Suriyah, M. R., Leibfried, T., Wössner, R., Weisser, K., and Frank, M., Eds. 2015. *Verification of a state of the art all-vanadium flow battery model with the "Compact Storage" by SCHMID Energy Systems GmbH*.

[11] Kangro, W. 1949. *Verfahren zur Speicherung von elektrischer Energie*, DE9144264C.

[12] Kangro, W. and Pieper, H. 1962. Zur Frage der Speicherung von elektrischer Energie in Flüssigkeiten. *Electrochimica Acta* 7, 4, 435–448.

[13] Skyllas-Kazacos, M. 1988. *All-vanadium redox battery*, US4786567A.

[14] Wandschneider, F. T., Finke, D., Grosjean, S., Fischer, P., Pinkwart, K., Tübke, J., and Nirschl, H. 2014. Model of a vanadium redox flow battery with an anion exchange membrane and a Larminie-correction. *Journal of Power Sources* 272, 436–447.

[15] Noack, J., Roznyatovskaya, N., Herr, T., and Fischer, P. 2015. The Chemistry of Redox-Flow Batteries. *Angewandte Chemie (International ed. in English)* 54, 34, 9776–9809.

[16] König, S., Suriyah, M. R., and Leibfried, T. 2016. Innovative model-based flow rate optimization for vanadium redox flow batteries. *Journal of Power Sources* 333, 134–144.

[17] Blanc, C. 2009. *Modeling of Vanadium Redox Flow Battery Electricity Storage System,* EPFL.

[18] Knehr, K. W., Agar, E., Dennison, C. R., Kalidindi, A. R., and Kumbur, E. C. 2012. A Transient Vanadium Flow Battery Model Incorporating Vanadium Crossover and Water Transport through the Membrane. *J. Electrochem. Soc.* 159, 9, A1446-A1459.

[19] Li, Y., Skyllas-Kazacos, M., and Bao, J. 2016. A dynamic plug flow reactor model for a vanadium redox flow battery cell. *Journal of Power Sources* 311, 57–67.

[20] Giorno, L., Strathmann, H., and Drioli, E. 2016. Mathematical Description of Mass Transport in Membranes. In *Encyclopedia of Membranes*, E. Drioli and L. Giorno, Eds. Springer Berlin Heidelberg, Berlin, Heidelberg, 1135–1138.

[21] Lei, Y., Zhang, B. W., Bai, B. F., and Zhao, T. S. 2015. A transient electrochemical model incorporating the Donnan effect for all-vanadium redox flow batteries. *Journal of Power Sources* 299, 202–211.

[22] Shah, A. A., Watt-Smith, M. J., and Walsh, F. C. 2008. A dynamic performance model for redox-flow batteries involving soluble species. *Electrochimica Acta* 53, 27, 8087–8100.

[23] Shah, A. A., Al-Fetlawi, H., and Walsh, F. C. 2010. Dynamic modelling of hydrogen evolution effects in the all-vanadium redox flow battery. *Electrochimica Acta* 55, 3, 1125–1139.

[24] Boettcher, P. A., Agar, E., Dennison, C. R., and Kumbur, E. C. 2015. Modeling of Ion Crossover in Vanadium Redox Flow Batteries: A Computationally-Efficient Lumped Parameter Approach for Extended Cycling. *J. Electrochem. Soc.* 163, 1, A5244-A5252.

[25] Knehr, K. W. and Kumbur, E. C. 2012. Role of convection and related effects on species crossover and capacity loss in vanadium redox flow batteries. *Electrochemistry Communications* 23, 76–79.

[26] Badrinarayanan, R., Zhao, J., Tseng, K. J., and Skyllas-Kazacos, M. 2014. Extended dynamic model for ion diffusion in all-vanadium redox flow battery including the effects of temperature and bulk electrolyte transfer. *Journal of Power Sources* 270, 576–586.

[27] Merei, G., Adler, S., Magnor, D., and Sauer, D. U. 2015. Multi-physics Model for the Aging Prediction of a Vanadium Redox Flow Battery System. *Electrochimica Acta* 174, 945–954.

[28] Tang, A., Bao, J., and Skyllas-Kazacos, M. 2011. Dynamic modelling of the effects of ion diffusion and side reactions on the capacity loss for vanadium redox flow battery. *Journal of Power Sources* 196, 24, 10737–10747.

[29] Tang, A., Bao, J., and Skyllas-Kazacos, M. 2012. Thermal modelling of battery configuration and self-discharge reactions in vanadium redox flow battery. *Journal of Power Sources* 216, 489–501.

[30] Tang, A., Bao, J., and Skyllas-Kazacos, M. 2014. Studies on pressure losses and flow rate optimization in vanadium redox flow battery. *Journal of Power Sources* 248, 154–162.

[31] Tang, A., McCann, J., Bao, J., and Skyllas-Kazacos, M. 2013. Investigation of the effect of shunt current on battery efficiency and stack temperature in vanadium redox flow battery. *Journal of Power Sources* 242, 349–356.

[32] Skyllas-Kazacos, M. and Goh, L. 2012. Modeling of vanadium ion diffusion across the ion exchange membrane in the vanadium redox battery. *Journal of Membrane Science* 399-400, 43–48.

[33] You, D., Zhang, H., Sun, C., and Ma, X. 2011. Simulation of the self-discharge process in vanadium redox flow battery. *Journal of Power Sources* 196, 3, 1578–1585.

[34] Kuhn, A. T. and Booth, J. S. 1980. Electrical leakage currents in bipolar cell stacks. *J Appl Electrochem* 10, 2, 233–237.

[35] Kaminski, E. A. 1983. A Technique for Calculating Shunt Leakage and Cell Currents in Bipolar Stacks Having Divided or Undivided Cells. *J. Electrochem. Soc.* 130, 5, 1103.

[36] White, R. E. 1986. Predicting Shunt Currents in Stacks of Bipolar Plate Cells. *J. Electrochem. Soc.* 133, 3, 485.

[37] Darling, R. M., Shiau, H.-S., Weber, A. Z., and Perry, M. L. 2017. The Relationship between Shunt Currents and Edge Corrosion in Flow Batteries. *J. Electrochem. Soc.* 164, 11, E3081-E3091.

[38] Skyllas-Kazacos, M., McCann, J., Li, Y., Bao, J., and Tang, A. 2016. The Mechanism and Modelling of Shunt Current in the Vanadium Redox Flow Battery. *ChemistrySelect* 1, 10, 2249–2256.

[39] Xing, F., Zhang, H., and Ma, X. 2011. Shunt current loss of the vanadium redox flow battery. *Journal of Power Sources* 196, 24, 10753–10757.

[40] Wandschneider, F. T., Röhm, S., Fischer, P., Pinkwart, K., Tübke, J., and Nirschl, H. 2014. A multi-stack simulation of shunt currents in vanadium redox flow batteries. *Journal of Power Sources* 261, 64–74.

[41] Corcuera, Sara and Skyllas-Kazacos, Maria. 2012. State-of-charge monitoring and electrolyte rebalancing methods for the vanadium redox flow battery. *European Chemical Bulletin* 1, 12, 511–519.

[42] Huggins, R. A. 2010. *Energy Storage*. Springer Science+Business Media LLC, Boston, MA.

[43] Mortimer, R. G. 2000. *Physical chemistry*. Harcourt/Academic Press, San Diego.

[44] Knehr, K. W. and Kumbur, E. C. 2011. Open circuit voltage of vanadium redox flow batteries: Discrepancy between models and experiments. *Electrochemistry Communications* 13, 4, 342–345.

[45] González-García, J., Bonete, P., Expósito, E., Montiel, V., Aldaz, A., and Torregrosa-Maciá, R. 1999. Characterization of a carbon felt electrode. Structural and physical properties. *J. Mater. Chem.* 9, 2, 419–426.

[46] Bromberger, K., Kaunert, J., and Smolinka, T. 2014. A Model for All-Vanadium Redox Flow Batteries. Introducing Electrode-Compression Effects on Voltage Losses and Hydraulics. *Energy Technology* 2, 1, 64–76.

[47] Zhou, H., Zhang, H., Zhao, P., and Yi, B. 2006. A comparative study of carbon felt and activated carbon based electrodes for sodium polysulfide/bromine redox flow battery. *Electrochimica Acta* 51, 28, 6304–6312.

[48] Al-Fetlawi, H., Shah, A. A., and Walsh, F. C. 2009. Non-isothermal modelling of the all-vanadium redox flow battery. *Electrochimica Acta* 55, 1, 78–89.

[49] Agar, E., Dennison, C. R., Knehr, K. W., and Kumbur, E. C. 2013. Identification of performance limiting electrode using asymmetric cell configuration in vanadium redox flow batteries. *Journal of Power Sources* 225, 89–94.

[50] Qiu, G., Joshi, A. S., Dennison, C. R., Knehr, K. W., Kumbur, E. C., and Sun, Y. 2012. 3-D pore-scale resolved model for coupled species/charge/fluid transport in a vanadium redox flow battery. *Electrochimica Acta* 64, 46–64.

[51] Chen, C. L., Yeoh, H. K., and Chakrabarti, M. H. 2014. An enhancement to Vynnycky's model for the all-vanadium redox flow battery. *Electrochimica Acta* 120, 167–179.

[52] Yang, W. W., He, Y. L., and Li, Y. S. 2015. Performance Modeling of a Vanadium Redox Flow Battery during Discharging. *Electrochimica Acta* 155, 279–287.

[53] Shah, A. A., Tangirala, R., Singh, R., Wills, R. G. A., and Walsh, F. C. 2011. A Dynamic Unit Cell Model for the All-Vanadium Flow Battery. *J. Electrochem. Soc.* 158, 6, A671.

[54] Sum, E. and Skyllas-Kazacos, M. 1985. A study of the V(II)/V(III) redox couple for redox flow cell applications. *Journal of Power Sources* 15, 2-3, 179–190.

[55] Yamamura, T., Watanabe, N., Yano, T., and Shiokawa, Y. 2005. Electron-Transfer Kinetics of Np^{3+}/Np^{4+}, NpO_2^+/NpO_2^{2+}, V^{2+}/V^{3+}, and VO^{2+}/VO_2^+ at Carbon Electrodes. *J. Electrochem. Soc.* 152, 4, A830.

[56] Gattrell, M., Park, J., MacDougall, B., Apte, J., McCarthy, S., and Wu, C. W. 2004. Study of the Mechanism of the Vanadium 4+/5+ Redox Reaction in Acidic Solutions. *J. Electrochem. Soc.* 151, 1, A123.

[57] You, D., Zhang, H., and Chen, J. 2009. A simple model for the vanadium redox battery. *Electrochimica Acta* 54, 27, 6827–6836.

[58] Fink, H., Friedl, J., and Stimming, U. 2016. Composition of the Electrode Determines Which Half-Cell's Rate Constant is Higher in a Vanadium Flow Battery. *J. Phys. Chem. C* 120, 29, 15893–15901.

[59] Friedl, J. and Stimming, U. 2017. Determining Electron Transfer Kinetics at Porous Electrodes. *Electrochimica Acta*.

[60] González, Z., Flox, C., Blanco, C., Granda, M., Morante, J. R., Menéndez, R., and Santamaría, R. 2016. Outstanding electrochemical performance of a graphene-modified graphite felt for vanadium redox flow battery application. *Journal of Power Sources*.

[61] Schmal, D., van Erkel, J., and van Duin, P. J. 1986. Mass transfer at carbon fibre electrodes. *J Appl Electrochem* 16, 3, 422–430.

[62] Delanghe, B., Tellier, S., and Astruc, M. 1990. Mass transfer to a carbon or graphite felt electrode. *Electrochimica Acta* 35, 9, 1369–1376.

[63] Kinoshita, K. 1982. Mass-Transfer Study of Carbon Felt, Flow-Through Electrode. *J. Electrochem. Soc.* 129, 9, 1993.

[64] Tjaden, B., Cooper, S. J., Brett, D. J. L., Kramer, D., and Shearing, P. R. 2016. On the origin and application of the Bruggeman correlation for analysing transport phenomena in electrochemical systems. *Current Opinion in Chemical Engineering* 12, 44–51.

[65] Zhou, X. L., Zhao, T. S., An, L., Zeng, Y. K., and Yan, X. H. 2015. A vanadium redox flow battery model incorporating the effect of ion concentrations on ion mobility. *Applied Energy* 158, 157–166.

[66] Zheng, Q., Zhang, H., Xing, F., Ma, X., Li, X., and Ning, G. 2014. A three-dimensional model for thermal analysis in a vanadium flow battery. *Applied Energy* 113, 1675–1685.

[67] Ma, X., Zhang, H., and Xing, F. 2011. A three-dimensional model for negative half cell of the vanadium redox flow battery. *Electrochimica Acta* 58, 238–246.

[68] Zheng, Q., Xing, F., Li, X., Ning, G., and Zhang, H. 2016. Flow field design and optimization based on the mass transport polarization regulation in a flow-through type vanadium flow battery. *Journal of Power Sources* 324, 402–411.

[69] Xu, Q., Zhao, T. S., and Leung, P. K. 2013. Numerical investigations of flow field designs for vanadium redox flow batteries. *Applied Energy* 105, 47–56.

[70] Wei, Z., Zhao, J., and Xiong, B. 2014. Dynamic electro-thermal modeling of all-vanadium redox flow battery with forced cooling strategies. *Applied Energy* 135, 1–10.

[71] Tomadakis, M. M. 2005. Viscous Permeability of Random Fiber Structures. Comparison of Electrical and Diffusional Estimates with Experimental and Analytical Results. *Journal of Composite Materials* 39, 2, 163–188.

[72] White, F. M. 2011. *Fluid mechanics*. McGraw-Hill, Singapore.

[73] Katz, J. 2013. *Introductory fluid mechanics*. Cambridge Univ. Press, Cambridge.

[74] Munson, B. R. 2010. *Fundamentals of fluid mechanics*. Wiley, Hoboken NJ u.a.

[75] RENNER GmbH. *Datasheet RM 1 - 5/35*.

[76] RENNER GmbH. *Datasheet RM 1.5 - 7/55*.

[77] RENNER GmbH. *Datasheet RM 3 - 10/120*.

[78] SIEBEC SAS. *Magnetic Drive Pumps M35 to M200 Range*.

[79] March Manufacturing Inc. *Datasheet TE-6T-MD*.

[80] You, X., Ye, Q., and Cheng, P. 2016. Scale-up of high power density redox flow batteries by introducing interdigitated flow fields. *International Communications in Heat and Mass Transfer* 75, 7–12.

[81] Kumar, S. and Jayanti, S. 2016. Effect of flow field on the performance of an all-vanadium redox flow battery. *Journal of Power Sources* 307, 782–787.

[82] Xu, Q., Zhao, T. S., and Zhang, C. 2014. Performance of a vanadium redox flow battery with and without flow fields. *Electrochimica Acta* 142, 61–67.

[83] Ke, X., Alexander, J. I. D., Prahl, J. M., and Savinell, R. F. 2014. Flow distribution and maximum current density studies in redox flow batteries with a single passage of the serpentine flow channel. *Journal of Power Sources* 270, 646–657.

[84] Knudsen, E., Albertus, P., Cho, K. T., Weber, A. Z., and Kojic, A. 2015. Flow simulation and analysis of high-power flow batteries. *Journal of Power Sources* 299, 617–628.

[85] Escudero-Gonzalez, J. and Lopez-Jimenez, P. A. 2014. Methodology to optimize fluid-dynamic design in a redox cell. *Journal of Power Sources* 251, 243–253.

[86] Prokopius, P. R. 1976. *Model for calculating electrolytic shunt path losses in large electrochemical energy conversion systems.* https://ntrs.nasa.gov/archive/nasa/casi.ntrs.nasa.gov/19760014614.pdf. Accessed 6 December 2016.

[87] Ye, Q., Hu, J., Cheng, P., and Ma, Z. 2015. Design trade-offs among shunt current, pumping loss and compactness in the piping system of a multi-stack vanadium flow battery. *Journal of Power Sources* 296, 352–364.

[88] Moro, F., Trovò, A., Bortolin, S., Del Col, D., and Guarnieri, M. 2017. An alternative low-loss stack topology for vanadium redox flow battery. Comparative assessment. *Journal of Power Sources* 340, 229–241.

[89] Yin, C., Guo, S., Fang, H., Liu, J., Li, Y., and Tang, H. 2015. Numerical and experimental studies of stack shunt current for vanadium redox flow battery. *Applied Energy* 151, 237–248.

[90] SGL-Group. *Speciality graphites for energy storage.* http://www.sglgroup.com/cms/_common/downloads/products/product-groups/gs/brochures/Specialty_Graphites_for_Energy_Storage_e.pdf. Accessed 2 February 2017.

[91] Schunk-Group. *FU4369 Data sheet.* http://stabiclim.com/sixcms/media.php/1722/FU_4369_en.pdf. Accessed 2 February 2017.

[92] SMA Solar Technology AG. *Sunny Island 6.0H / 8.0H for off-grid and on-grid applications.*

[93] TRUMPF Hüttinger. *Product website TruConvert.* https://www.trumpf.com/en_GB/products/powerelectronics/energy-storage-system/truconvert/?LS=1. Accessed 6 June 2017.

[94] Hagedorn, N. H., Hoberecht, M. A., and Thaller, L. H. 1982. *NASA Redox Cell Stack Shunt Current, Pumping Power, and Cell Performance Tradeoffs.* NASA.

[95] Blanc, C. and Rufer, A. Optimization of the operating point of a vanadium redox flow battery. *2009 IEEE Energy Conversion Congress and Exposition.*

[96] Xiong, B., Zhao, J., Tseng, K. J., Skyllas-Kazacos, M., Lim, T. M., and Zhang, Y. 2013. Thermal hydraulic behavior and efficiency analysis of an all-vanadium redox flow battery. *Journal of Power Sources* 242, 314–324.

[97] Ma, X., Zhang, H., Sun, C., Zou, Y., and Zhang, T. 2012. An optimal strategy of electrolyte flow rate for vanadium redox flow battery. *Journal of Power Sources* 203, 153–158.

[98] Ling, C. Y., Cao, H., Chng, M. L., Han, M., and Birgersson, E. 2015. Pulsating electrolyte flow in a full vanadium redox battery. *Journal of Power Sources* 294, 305–311.

[99] Forsythe, G. E., Malcolm, M. A., and Moler, C. B. 1977. *Computer methods for mathematical computations*. Prentice-Hall series in automatic computation. Prentice-Hall, Englewood Cliffs NJ.

[100] König, S., Suriyah, M. R., and Leibfried, T. 2017. A plug flow reactor model of a vanadium redox flow battery considering the conductive current collectors. accepted - in press. *Journal of Power Sources*.

[101] AMG Titanium Alloys & Coatings. *Vanadium Electrolyte Solution 1.6M.*

[102] Sun, C., Chen, J., Zhang, H., Han, X., and Luo, Q. 2010. Investigations on transfer of water and vanadium ions across Nafion membrane in an operating vanadium redox flow battery. *Journal of Power Sources* 195, 3, 890–897.

[103] Tang, A., Ting, S., Bao, J., and Skyllas-Kazacos, M. 2012. Thermal modelling and simulation of the all-vanadium redox flow battery. *Journal of Power Sources* 203, 165–176.

Appendix E

Author's publication and supervised student theses

E.1 Author's publication in chronological sequence

Peer-reviewed journal publications

S. König, M. Suriyah and T. Leibfried
Model based examination on influence of stack series connection and pipe diameters on efficiency of vanadium redox flow batteries under considerations of shunt currents
Journal of Power Sources 281 (2015), pages 272-284
DOI: 10.1016/j.jpowsour.2015.01.119

S. König, M. Suriyah and T. Leibfried
Innovative model-based flow rate optimization for vanadium redox flow batteries
Journal of Power Sources 333 (2016), pages 134-144
DOI: 10.1016/j.jpowsour.2016.09.147

M. Uhrig, S. König, M. Suriyah and T. Leibfried
Lithium-based vs. Vanadium Redox Flow Storage Batteries – A comparison for Home Storage Systems
Energy Procedia 99 (2016), pages 25-43
DOI: 10.1016/j.egypro.2016.10.095

S. König, M. Suriyah and T. Leibfried
A plug flow reactor model of a vanadium redox flow battery considering the conductive current collectors
Journal of Power Sources 360 (2017), pages 221-231
DOI: 10.1016/j.jpowsour.2017.05.085

Publications in conference proceedings

S. König, P. Müller and T. Leibfried
Design and comparison of three different possibilities to connect a vanadium-redox-flow-battery to a wind power plant
13th Spanish-Portuguese conference on electrical engineering, 2013, Valencia, Spain

S. König, M. Zimmerlin, T. Leibfried, F. Wandschneider, S. Röhm and P. Fischer
Optimaler Aufbau von Batteriesystemen für Windkraftanlagen
VDE ETG Congress 2013, Berlin, Germany

S. König and T. Leibfried
Modellierung und Simulation eines 2 MW/20 MWh Vanadium Redox Flow Batteriesystems
NEIS Conference 2013, Hamburg, Germany

S. König, M. Suriyah and T. Leibfried,
Optimal channel design for minimizing shunt currents in all-vanadium redox flow batteries
Advanced Automotive Battery Conference (AABC Europe), 2015, Mainz, Germany

S. König, K. Iliopoulos, M. Suriyah and T. Leibfried
How direct marketing of a Wind Power Plant affects optimal Design and Operation of Vanadium Redox Flow Batteries
9th International Renewable Energy Storage Conference (IRES), 2015, Düsseldorf, Germany

S. König, M. Suriyah and T. Leibfried
Volumetric Flow Rate Control in Vanadium Redox Flow Batteries Using a Variable Flow Factor
Best Paper Award
International Renewable Energy Congress (IREC), 2015, Sousse, Tunesia
DOI: 10.1109/IREC.2015.7110861

S. König, A. Wöber, M. Suriyah, T. Leibfried, R. Wößner, K. Weisser and M. Frank
Verification of a state of the art all-vanadium flow battery model with the "Compact Storage" by SCHMID Energy Systems GmbH
International VDE ETG Congress, 2015, Bonn, Germany
Available on IEEEXplore, ISBN: 978-3-8007-4121-2

S. König, Q. Bchini, R. McKenna, W. Köppel, M. Bachseitz and J. Michaelis
Spatially-resolved analysis of the challenges and opportunities for Power-to-Gas (PtG) in Baden-Württemberg until 2040
11th International Renewable Energy Storage Conference (IRES), 2017, Düsseldorf, Germany

S. König, M. Uhrig, D. Suriyah, T. Leibfried, R. Höche, M. Schroeder, H. Buschmann and R. Wößner
Hybrid-Optimal – Demonstrating the cellular approach with a hybrid battery concept
11th International Renewable Energy Storage Conference (IRES), 2017, Düsseldorf, Germany

Oral presentations without paper

S. König and T. Leibfried
Holistic Modeling and Optimization of Vanadium Redox Flow Batteries
Advanced Battery Power 2014, Münster, Germany

M. Uhrig, S. König, M. Suriyah and T. Leibfried
Lithium-based vs. Vanadium Redox Flow Storage Batteries – A comparison for Home Storage Systems
10th International Renewable Energy Storage Conference (IRES), 2016, Düsseldorf, Germany

S. König, M. Uhrig and T. Leibfried
Redox flow batteries – An alternative for home storage systems?
Advanced Battery Power, 2016, Münster, Germany

S. König, M. Uhrig and T. Leibfried
Economics of the Vanadium Redox Flow Battery for home- and community storage
International Flow Battery Forum (IFBF), 2016, Karlsruhe, Germany

Poster presentations without paper

S. König and T. Leibfried
Introduction of a Flow Battery Management System (FBMS)
Advanced Battery Power, 2015, Aachen, Germany

S. König and T. Leibfried
Optimal flow rate and load sharing control for flow batteries
Energy, Science & Technology (EST), 2015, Karlsruhe, Germany

S. König, M. Suriyah, T. Leibfried, T. Lüth, D. Leypold and J. Bornwasser
Optimal system design for all-vanadium redox flow batteries considering the power-electronic grid connection
International Flow Battery Forum (IFBF), 2015, Glasgow, UK

S. König, M. Suriyah and T. Leibfried
An innovative approach for the model-based flow rate optimization of vanadium redox flow batteries
International Flow Battery Forum (IFBF), 2016, Karlsruhe, Germany

S. König, T. Leibfried, M. Schroeder and P. Schauer
Vanadium Redox Flow Batteries – Model vs. Reality
Batterieforum Deutschland, 2017, Berlin, Germany

E.2 Supervised Student Theses

Master and diploma theses

P. Müller
Systemanalyse einer batteriegestützten Windkraftanlage - 2012

K. Iliopoulos
Erstellung und Simulation von Batteriefahrplänen zur Frequenzregelung und zur Direktvermarktung von Windenergie - 2014

K. Stumm
Identifikation und Simulation von Möglichkeiten zur Unterdrückung von Streuströmen in großen Vanadium-Redox-Flow-Batteriesystemen - 2014

A. Wöber
Erweiterung eines multiphysikalischen Modells für Vanadium-Redox-Flow-Batterien - 2015

Bachelor theses / Seminar papers

T. Zimmermann
Entwurf und Aufbau einer Teslaspule mit Funkenstrecke - 2012

E. Elcheroth
Entwurf und Aufbau einer Teslaspule mit einer Elektronenröhre - 2012

J. Prieto
Modellierung einer Windkraftanlage mit Ansoft Simplorer - 2012

N. Goldbeck
Modellierung einer Windkraftanlage mit Matlab/Simulink - 2012

E. Freer
Entwicklung und Aufbau einer drehzahlgeregelten Ansteuerung eines permanent erregten Gleichstrommotors - 2013

M. Backes
Simulative Bestimmung von Verlusten in leistungselektronischen Schaltungen - 2013

A. Glatt
Untersuchung des Potentials von Nachtspeicherheizungen zur gezielten Aufnahme regenerative erzeugter Energie - 2013

M. Zimmerlin
Automatisierte Modellierung von Vanadium-Redox-Flow-Batterien – 2013

F. Zghal
Vergleich von Umrichtertopologien zur Anbindung von großen Batteriesystemen an das Netz – 2014

H. Krämer
Simulation und Auslegung einer Power-to-Gas Anlage - 2014

K. König
Multiphysikalische Simulation von Vanadium-Redox-Flow-Batteriezellen zur Bestimmung von Druckverlusten und Streuströmen - 2014

S. Paroth
Verifikation der Verlustsimulation leistungselektronischer Schaltungen an Prototypen - 2014

E. Eßwein
Entwicklung eines Betriebsmanagements (BMS) für große Vanadium-Redox-Flow-Batteriesysteme - 2015